Robert Maillart's Bridges

Princeton University Press   New Jersey

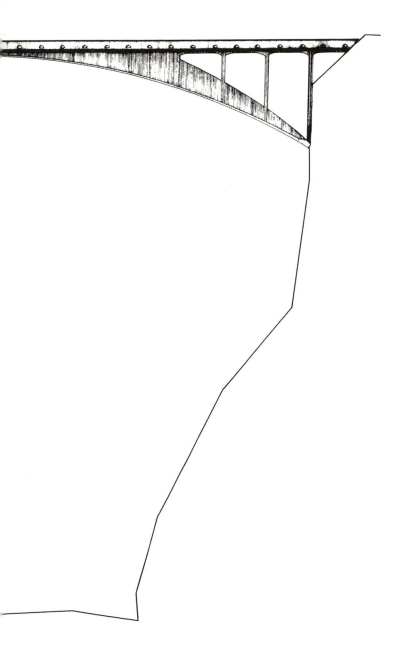

ROBERT MAILLART'S BRIDGES
THE ART OF ENGINEERING

David P. Billington

TG
140
.M3
B55

Contents

Illustrations and Tables

Illustrations

Unless otherwise stated, photographs are by the author, and diagrams are by A. Evans or T. Agans.

Prologue

Chapter 1

Chapter 2

Chapter 3

Chapter 4

Chapter 5

Tables

Chapter 4

Chapter 9

Preface

This study grew out of a desire to connect the visual elegance of Robert Maillart's works to their technical basis. The book, therefore, centers on a small number of the most significant works and on the major ideas of Robert Maillart; its primary goal is to explore structural form as it arises out of aesthetic feelings and scientific ideas.

I meant this work for the general reader who is interested in bridges or in other engineering structures that have the potential for highly expressive visual form. I have sought to write a text that would be widely understood yet consistent with the terminology of the engineer or architect. Because some of the argument depends upon mathematical formulations, the notes contain numerous calculations and technical explanations, which those with such interests can pursue. My original intention was to write a brief biography. However, the subsequent discovery of a rich supply of private letters and business correspondence has made a full biography possible, but only after much more study. I have started work on it and plan to complete it in the near future.

This work would have been impossible without the continual encouragement, support, and substantial collaboration of Robert Maillart's daughter, Marie-Claire Blumer-Maillart, and her husband Edouard Blumer. They have worked tirelessly to preserve, collect, and help document the papers, the works, and the events in the life of Robert Maillart. That I could seriously take up the long-term study of Maillart is primarily due to their devotion, both to him and to the preservation of his documents. René Maillart was also of great help in giving me his own perspective on his father.

I am also deeply indebted to Professor Christian Menn, one of the foremost contemporary bridge designers, whose clear understanding of Maillart's works has guided much of my efforts. He has also put me in touch with local cantonal and city engineers from whom I obtained many essential documents. In particular, I want to thank Engineers Letta and Bosch of St. Gallen, Engineers Stampf and Tschudin of Chur, Engineer Schlumpf of the Rhätische Bahn, and Engineer Hirt and Mr. Stalden of Zurich.

The firms that took over Maillart's offices in Geneva and Bern have made their records available to me. Engineer Tremblet in Geneva has allowed me free access to the substantial Geneva archive of Maillart's works; he also gave his time to have documents reproduced and sent to Princeton, while Princeton University Library kindly bore most of the costs of reproducing them. Engineer Bernet in Bern has also freely given his time to sort documents and to send copies to Princeton.

The most valuable professional recollections of Maillart have come to me through the kindness of Ernst Stettler, Maillart's chief engineer in Bern from 1926 to 1940 and his successor there until 1962, when he turned the office over to Bernet and Weyenroth. The architect Hans Kruck has shared his invaluable understanding of Maillart from the perspective of art and creativity, and Marcel Fornerod has provided a firsthand account of his two years in Maillart's Zurich office. Many more Swiss have helped me in my work and made my association with their country a happy personal one. As with anyone studying Maillart, I am indebted to Sigfried Giedion and Max Bill, whose pioneering writings showed the beauty of his works to a wide audience.

Basic financial support for this study came from a long-term grant from the National Endowment for the Humanities and I would like to record with special gratitude the encouragement of Herbert McArthur, former program officer there. The Ford Foundation provided substantial matching funds, and a grant from the Rockefeller Foundation helped with the initial research. This research accelerated greatly after a conference, held at Princeton in 1972, commemorating the centennial of Maillart's birth; this was also supported by the National Endowment for the Humanities. My work in Switzerland was materially furthered by a grant from the Swiss Society of Cement, Chalk, and Gypsum Manufacturers whose executive officer, Dr. H. Eichenberger, provided not only funds but considerable aid in documentation and insight into Swiss culture.

Mrs. Betty Mate deserves a separate paragraph for her good will, her endurance, and her consummate skill in typing the first complete manuscript and persevering through all the preliminary workings. As a superlative secretary, she both protected me and kept me in line, and I am grateful. I appreciate also the help of Jeanne Carlucci with the manuscript. When Mrs. Mate left Princeton, I was lucky enough to find another excellent secretary, Thelma Keith, who typed another version with patience and skill following the fine editing of Ruth Bonner. I am indebted to Edward Tenner of Princeton University Press for numerous helpful suggestions, to Judith May for her sympathetic editorial work, and to Etta Recke for her thoughtful and efficient typing of the final result. An additional pleasure has been the lively and useful work on Maillart by a series of students at Princeton; those who have contributed directly to the research for this book are Michael Hein, Robert Shulock, David Lamb, Ellen Leing, Mark Herron, Kent Smith, Neil Hauck, Howard Miller, and a special thanks to

James Chiu. Further, I acknowledge with thanks the help of my colleagues John Abel and Robert Mark who have done Maillart research themselves and have read critically much of my early writing on Maillart.

Finally this work is, in reality, a family effort in far more than the ordinary sacrificial way that wives and children suffer during the writing of a new book. My wife, Phyllis, and my three eldest children, David, Elizabeth, and Jane worked actively with me by inspecting bridges, collecting archival material, and reading initial drafts. In particular, my son David has read carefully the entire manuscript and suggested major re-writing to the end that the nonengineer can follow my arguments. His keen editorial judgment is reflected throughout.

David P. Billington
April 7, 1977

Robert Maillart's Bridges

Prologue   The Salgina Crossing

Traveling out from the ordered wealth of Zurich, the train passes two mountain lakes, the Zurichsee and the Wallensee, before crossing the Rhine River at Bad Ragaz. Continuing south toward Chur, capital of the Swiss Canton of the Graubünden, the train stops at the small town of Landquart. Here one can change trains and take the colorful, privately owned narrow-gauge Rhätische Bahn up the valley of the Landquart River to the village of Schiers, where the Schraubach stream joins the Landquart on its way to the Rhine. From Schiers a single-lane road winds up to the mountain village of Fajauna. Few signs of civilization can be seen above the high Alpine meadows as the road curves up the southern slope of the Rätikon range that separates Switzerland from Austria. After one curve, a small white form appears through the trees. After a few more curves, it comes into full view—a bridge, connecting two mountains over a wild ravine. To laymen its form is unclear at first and then distinguishable as a bridge. To knowledgeable engineers, however, it is not only immediately clear, it is also the reason for the pilgrimage. Here is one of the most beautiful examples of pure twentieth-century structure. But it is also complex and, even to the skilled engineer, an object of mystery and wonder.

This bridge, the Salginatobel, was completed in 1930 to serve the population of Schuders, an Alpine community of less than fifty people; yet it was the focus of the first art museum exhibition ever devoted to pure engineering, held at New York City's Museum of Modern Art in 1947. Designed by a highly trained engineer whose highest school grade was in differential equations, the bridge calculations employed elementary mathematics with no calculus at all. Expressing in its form one of the simplest of all technical ideas, it nevertheless had practically no precedent. Arising in a setting of almost primitive mountain simplicity, the bridge almost immediately became an important force in the most sophisticated circles of avant-garde intellectuals. Considered now to be a work of art, the design was originally chosen because it was the least expensive proposal.

This work of pure engineering strikes laymen and engineers alike as something radically new and prototypical of the twentieth century. Its designer, Robert Maillart, created in the wilderness a bridge of such extraordinary beauty that its material, reinforced concrete, became the medium for a legitimate style in its own right.

Robert Maillart through the Salginatobel focused the meaning of modern technology, not just on constructional possibilities but also on the potential inherent in designs that connect the scientific concept of efficiency to the artistic vision of elegance, under the twentieth-century tenet that the one supports the other. This bridge exemplifies the ideal that in the modern structuring of an environment, efficiency and elegance are merely aspects of the same design seen from the perspectives of science and of art; and that the essence of engineering lies in the integration of the two by the connecting link of economy. Maillart's design of the Salginatobel marked the beginning of his most creative period, which lasted until his death ten years later.

I have described a trip up the ancient mountains of Switzerland and into modern concepts of art and technology as an introduction to the study of a few of Maillart's significant works and of several of his central ideas. These works and ideas will form the basis of an evaluation of Maillart as a great engineer whose cultural significance parallels that of the great scientists or painters of our century.

1 Family and School (1872-1894)

Robert Maillart was born on February 6, 1872 of a long line of professionals, businessmen, and artists. The family was originally from Belgium and descended from a certain Jean Coley dit Maillart, who was recognized in 988 for valor in battle under the Bishop of Liège. His loss of sight on the battlefield is supposed to have been the origin of the French version of Blind Man's Buff, which is known in French as "Colin-Maillard."[1]

Robert Maillart's great-grandfather, Philippe Joseph Maillart (1764-1856), had a distinguished career, first as an engraver and later as a landscape painter.[2] His works are to be found in Belgium, and his landscapes still belonging to the Swiss Maillart family depict charming scenes of small houses and countrysides. Maillart's grandfather, Hector, married a Protestant, Petronelle Hubertine Schirmer, in 1833. They eventually settled at Carouge, near Geneva, in 1852 and Hector acquired Swiss citizenship in 1858.[3] The older of their two children, Edmond (1834-1874), was Maillart's father, about whom very little is known.[4] In 1866, he married Bertha Küpfer (1843-1932), the daughter of a prominent family of Bern. Edmond and Bertha Maillart had six children, of whom Robert was the fifth.[5] When Robert was two, his father died, leaving the family with little money.

From 1885 until 1889, Robert Maillart attended the Bern Gymnasium, where he excelled in mathematics and drawing, both artistic and technical. In 1889, he passed his state examinations with a grade average of 4.8 out of 6.0. His best subjects were descriptive geometry and his weakest, language and natural history.[6] At seventeen, he qualified for admission to the Swiss Federal Technical University, the Eidgenössische Technische Hochschule (ETH), in Zurich. But since the entrance age had been fixed by law at eighteen, Maillart spent the next year at the municipal school of watchmaking in Geneva. Finally, in October 1890, he entered the ETH, then as now one of the finest technical institutes in Europe.

Founded in 1855 as the Polytechnische Schule, the institute began with five departments; architecture, civil engineering, mechanical engineering, chemical engineering, and forestry, all under a corps of professors chosen from Switzerland and neighboring countries. For Maillart's education, the two most important initial appointments were the Germans Gottfried Semper (1803-1879) and Carl Culmann (1821-1881), each probably the most influential European academic in his field during the middle of the nineteenth century.

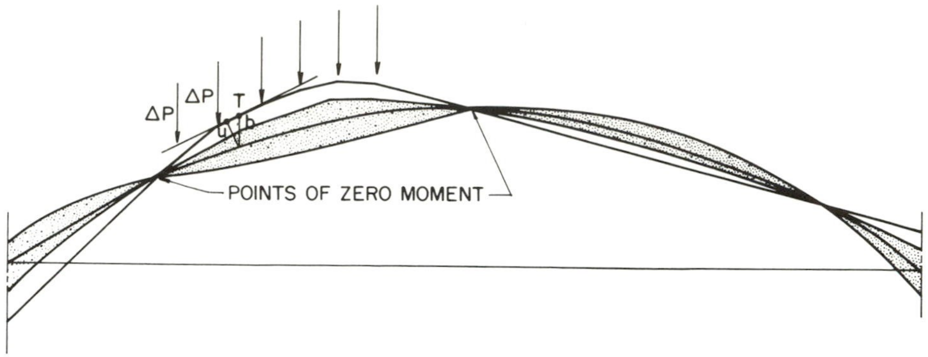

POINTS OF ZERO MOMENT

Fig. 1-1a. Diagram taken from *Graphic Statics* by Carl Culmann, 2nd ed., 1875, p. 594.

Semper had been a practicing architect and director of the School of Architecture in Dresden from 1834 to 1849, when he had to leave Germany following the abortive revolution. His completed works and his writings exerted a strong influence,[7] promoting the "rationalist" or structural approach to architecture in German-speaking Europe. Called to the ETH from London in 1855 at the suggestion of Richard Wagner, who was then living in Zurich, Semper was the first professor appointed in the architecture department (then called the building school). He left his imprint not only by designing the main Institute building, but also through his ideas.[8]

Maillart, coming to the ETH after Semper's death, absorbed this influence largely in his courses on building construction under the architect Benjamin Recordon (1845-1938), who had been appointed professor in the architecture department in 1890. In class notes that Maillart kept to the end of his life there appear some beautiful sketches of structures about which Recordon had lectured. Engineering students at the ETH continued to be taught by visually oriented architects within the school founded by Semper.[9]

Even more important was the influence of Carl Culmann of the Rhenish Palatinate, the first appointment made in the Institute's department of civil engineering.[10] Culmann had studied engineering at the Polytechnical School in Karlsruhe, had worked in building and bridge construction in Germany, and had traveled for two years through the British Isles and the United States studying bridges, railroads, and steamship construction. Although well trained in mathematics, Culmann brought to Zurich in 1855 the idea that structural calculations could be made graphically, and immediately began writing his great work *Graphic Statics*, which strongly influenced all engineering education and practice for the next half century. It is interesting to compare Culmann's arch moment diagram (from the 1875 edition, Fig. 1-1a) with

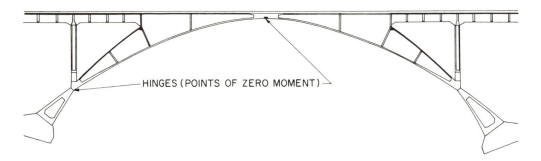

HINGES (POINTS OF ZERO MOMENT)

Fig. 1-1b. Maillart's design for the railroad bridge competition, Bern, 1935

Maillart's design for the Bern railway bridge competition sixty years later (Fig. 1-1b). The visually suggestive nature of Culmann's diagram indicates the design potential inherent in his methods.[11] While there is no evidence that Maillart copied this diagram for his Bern design, he did learn at the ETH the habit of connecting force diagrams to design forms.

When Culmann died in 1881, he was replaced by his student Wilhelm Ritter,[12] who not only took over his courses on the graphical analysis of structures and on bridge design, but also continued his work in graphic statics. Maillart took Ritter's courses, and his class notes clearly show his early interest both in the graphic analysis of bridges and in the wide variety of bridge forms described by Ritter, who like Culmann had traveled to the United States for study.[13]

Ritter used to end lectures on wooden bridges with a brief section on arch bridges, in which he named four systems, each described essentially in terms of stiffening. Maillart's class notes reveal not only Ritter's ideas but also something of Maillart's own reactions: Swiss bridges were "very complicated"; "to be noted [were] the elegant American examples of braced truss work"; and finally at the very end, after writing down Ritter's description of the fourth system—which is the only one that could be called fully deck-stiffened—Maillart added the note, "marvelous bridge."[14]

Maillart saved these class notes and a few others and, like most Swiss structural engineers of his generation, was influenced by Wilhelm Ritter all his life. In Maillart's first set of calculations for a deck-stiffened arch, his basic computations contained only one reference, which read as follows: "See Prof. W. Ritter. Collections of the complete technical writings, under stiffening beams."[15] The technical foundation of deck-stiffened arches, is to a large extent, the work of Ritter.

The ETH civil engineering curriculum of 1890-1894 was far more visually oriented than that of engineering schools in the mid-twentieth century, and probably more so than that of any other school in the 1890's. It is perhaps no accident that the two most outstanding bridge designers in the first half of the twentieth century—one using concrete (Maillart) and the other steel (Othmar Ammann)—were both graduates of the ETH, where they had studied under essentially the same faculty.[16]

Maillart's academic record shows him to have been proficient in mathematics and drawing, poor in theoretical machine studies, and only fair in bridges and the theory of structures. His highest grades were in the theory of ordinary differential equations and descriptive geometry. When he received his diploma on March 17, 1894, he was probably as well trained as any engineering student in Europe; in fact, the only two courses in basic civil engineering taught today that Maillart did not have then were on reinforced concrete and statically-indeterminate structural analysis, just the two areas in which he was to make revolutionary contributions. To appreciate the nature of his work, it is essential first to understand the properties of reinforced concrete and its development up to the 1890's when Robert Maillart began his career.

2 Reinforced Concrete in the 1890's

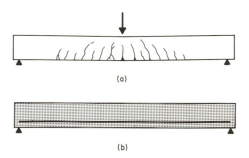

Concrete is a building material consisting of gravel or crushed stone, sand, cement, and water. Because it is a fluid mass when mixed, it can be cast into forms of any shape desired before hardening. Reinforced concrete is made by casting the concrete over a cage of steel bars. An unreinforced concrete beam will bend downward under its own weight, especially at the center, causing the top to compress together but stretching the bottom apart in tension. Concrete carries compression easily, but under very low tension it cracks (Fig. 2-1a). However, if reinforcing steel bars are embedded along its bottom inside, the bars can carry tension while the concrete in the top part of the beam can carry the compression (Fig. 2-1b).

When Robert Maillart graduated from the ETH in 1894, reinforced concrete was little used in Switzerland. The ETH taught no course in it, and no Swiss engineer had ventured to design a major structure in such a material.[1] Masonry, wood, and metal were still the basic materials of engineering construction, as they had been since the Roman Empire, although unreinforced concrete was known even then.

Joseph Monier, a French gardener, was one of the first to think of reinforcing concrete. Around 1867 he strengthened concrete tanks and pipes by casting the concrete over a skeleton of iron, and patented the idea. But lacking technical training, Monier never fully realized *why* such reinforcing worked and so was partly unaware of engineering applications for his invention. It was a civil engineer, G. A. Wayss of Berlin, who recognized the potential of reinforced concrete as a large-scale building material when he saw some of Monier's work at the Antwerp Exhibition of 1885. After purchasing the patent rights from Monier, Wayss established in Berlin a "Corporation for Concrete and Monier Construction," which built reinforced concrete structures all over Germany in the 1880's. Sadly, Monier himself gained nothing from Wayss's subsequent success, having sold the rights for a block sum instead of for a royalty.[2]

The use of reinforced concrete became widespread only after 1894, however, and the leading European figure, along with Wayss, was François Hennebique (1842-1921). Beginning as a stone mason, Hennebique had established a construction business in France that included the restoration of gothic cathedrals. In 1880, he launched a twelve-year program of secret research, which led to the development of

Figs. 2-1a and 1b. (a) Tension cracks in a gravity-loaded concrete beam (b) Reinforcing bars to carry the tension in a concrete beam

a complete system of design and construction in reinforced concrete similar to, but independent of, the system developed by Wayss. In 1892, Hennebique took out patents on his method, retired from construction, and organized an international system of licensees to carry out his designs. By 1917, Hennebique's firm had completed over 17,000 building contracts and about the same number of engineering designs in reinforced concrete.[3]

In 1902 Paul Christophe, a former member of Hennebique's staff, published *Reinforced Concrete and Its Applications*, in which he gave a detailed picture of the uses of reinforced concrete ten years after Hennebique's method was patented.[4] The book presented hundreds of photographs and diagrams of all kinds of structures made of reinforced concrete: multistory buildings (Fig. 2-2), beam and arch bridges, pipes, water towers, thin-shell domes, barrel vaults, and canals. Christophe provided a description of the many systems that had arisen in competition with Wayss and Hennebique, and also gave a detailed derivation of the formulas used at the time to determine concrete stresses and the amounts of steel consequently needed as reinforcement.

In the United States, progress in reinforced concrete building from 1894 to 1904 was summarized at an international engineering congress held in St. Louis in 1904.[5] Sponsored by the American Society of Civil Engineers (ASCE), the congress was intended to be a "review of progress during the past decade" in thirty-seven subjects, one of which was "concrete and concrete-steel," i.e., reinforced concrete.

An opening paper noted that before 1894, no important concrete-steel bridge had been built in the United States. Comparing the advantages of reinforced concrete bridges over plain steel bridges, the author continued: "They make handsome structures, and architectural ornament can be applied to any extent desired; if properly designed and constructed, they have vastly greater durability and greater ultimate economy; they are comparatively free from vibration [shaking] and noises; they are proof against tornadoes and fire, and also against floods, if the foundations are protected from scour; the cost of maintenance is confined to the pavements."[6] To all these aesthetic and structural advantages, he added one further characteristic of reinforced concrete that would prove crucial to its use in Switzerland: extensive use of local labor and local material keeps the flow of money in the community that pays for them. As cantonal autonomy is an important Swiss political priority, this domestic character of concrete made it much more appealing, especially in remote Alpine districts, than competing building materials such as steel, which had to be imported over great distances.

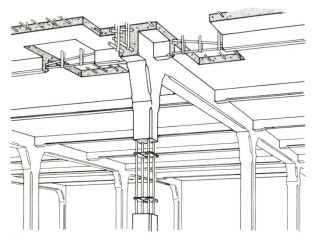

Fig. 2-2. Hennebique system for multistory reinforced-concrete buildings

Other papers revealed reinforced concrete to be an international development that had grown in popularity throughout the West. As Dr. Fritz von Emperger of Austria declared, "Ten years ago, the number of concrete-steel bridges was so small that there would have been no difficulty in giving a complete list, whereas now it would be quite impossible."[7] He did note the number of Hennebique bridges, which went from 5 in 1894 to 109 in 1903! Von Emperger emphasized the system invented by his fellow Austrian, Josef Melan, who built a steel arch bridge first, and then, hanging wooden forms from the steel, cast concrete around the steel arch, thereby getting a reinforced concrete arch without having to build scaffolding.

At the same time, von Emperger reviewed in some detail numerous other concrete arches, including the longest Hennebique arch bridge, crossing the Vienne River at Châtellerault, France, with three spans of 40 meters, 50 meters, and 40 meters.[8] He observed that the entire structure was one piece without any joints or hinges, and for this reason it showed a number of cracks. He further explained that when cracks in concrete occurred, the cracked sections can rotate somewhat, as if they had hinges. He concluded that building three hinges into a concrete arch would eliminate the cracking by allowing the arch to expand or contract freely under temperature change or small settlements in the foundations, without adding any stresses to the materials. As shown in Figure 2-3a, a hingeless arch subjected to a temperature rise will crack near both abutments and near its crown, whereas a three-hinged arch (Fig. 2-3b) can expand freely without cracking due to the rotation at the hinges. Early hinges in concrete arches consisted literally of metal hinges embedded in the concrete (Fig. 2-4). Maillart would follow the practice of hinging the arches in his early bridges, and the heavily cracked Hennebique bridge undoubtedly influenced his designs.

The "*Système Hennebique*" had been patented in Switzerland in early 1893 and its advantages were described in a brief article in the *Schweizerische Bauzeitung* (*SBZ*) in 1895. Further articles appeared in 1897.[9] In 1899 Wilhelm Ritter published a series of detailed articles describing the Hennebique system, showing how to make the basic calculations, and commenting briefly on the applications possible.[10] Ritter's work was really a brief text on the general principles of reinforced concrete design, because almost everything he wrote addressed the basic problems rather than those peculiar to one of the many different systems then prevalent. Written at the same time as Christophe's book (which had first appeared as a series of magazine articles in 1899), Ritter's articles helped to win acceptance for reinforced concrete in Switzerland. Of particular importance were his calculation examples, which clearly

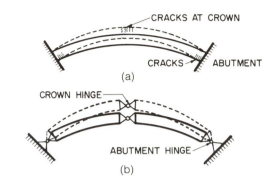

Figs. 2-3a and 3b. (a) Tension cracks in a heated concrete arch (b) Free rotation without cracks in a heated three-hinged concrete arch

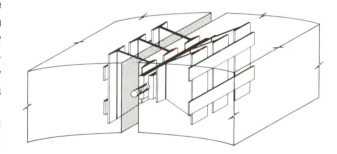

Fig. 2-4. Typical metal hinge used in concrete arch bridges

Fig. 2-5. Coal silo for the Aniche mines by Hennebique

showed any trained engineer how to determine appropriate concrete dimensions and amounts of reinforcement.

For Maillart's later work, two aspects of Ritter's exposition were significant. First was his brief description of how an arched beam (*Bogenträger*) in reinforced concrete could be studied easily by principles well known to any of his former students. It was just this type of structure that Maillart would develop in his first major new idea for a bridge, at Zuoz over the Inn River. The second important aspect of Ritter's articles was his skepticism about long-term adhesion between steel and concrete, particularly under vibrations or temperature changes. He did not believe that engineers had had enough experience with reinforced concrete to be able to state conclusively that the steel bars would stay bonded to the concrete over the lifetime of a structure.

This doubt provoked the engineer S. de Mollins, the general agent for Hennebique patents in Switzerland, to defend the validity of the system in a letter to the *SBZ*.[11] De Mollins emphasized that while Hennebique had built his first large floor in reinforced concrete as early as 1879 and his second in 1889, he had studied them carefully and waited until 1892 before publicizing his system. This period had given Hennebique the experience necessary to design an extraordinarily wide variety of structures. According to de Mollins, adhesion had been proven effective and, for reinforced concrete, all basic problems had been solved.

Robert Maillart was impressed enough with de Mollins's arguments to write an article of his own following a lecture to the Zurich branch of the Swiss Society of Engineers and Architects (SIA) by de Mollins in 1901. He described eight applications of reinforced concrete out of the great many shown by de Mollins. The article was enthusiastic and uncritical, suggesting that its author was beginning to see great opportunities opening up in a new field in which he was well prepared to work. The Hennebique works shown by de Mollins demonstrated for Maillart that the *Grand Prix* that Hennebique had received in the Paris Exposition of 1900 was well deserved.[12]

The eight applications can be taken as a brief summary of the state of concrete construction at the turn of the century just before Maillart began to create his own original works. Two of the examples were factory shed roofs of which the light-framed structures were fully visible from within, creating an unusually airy impression, according to Maillart. The third example was a widespanning arch roof that showed that reinforced concrete could compete with steel even there; and the next two were from interiors in Genoa showing how reinforced concrete could be used for

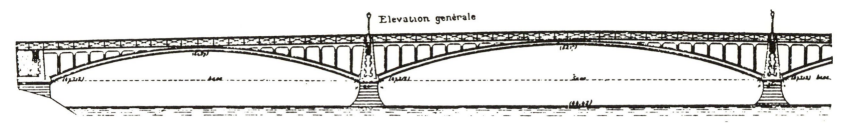

Elevation générale

decorated interiors with columns, arcades,and beams. It was, however, the last three that produced Maillart's greatest expressions of enthusiasm. Of application six (Fig. 2-5), a coal silo, Maillart remarked that to someone unfamiliar with reinforced concrete this would look very peculiar. But for him it was a "daring demonstration of frame construction built without diagonals and able to withstand all loadings with barely the slightest movement."

After briefly describing an elegantly built elevated water tank as the seventh example, he finally discussed the bridge at Châtellerault (Fig. 2-6) which he called "the most important engineering work to date built out of reinforced concrete." He noted its pleasing appearance and its safety: it had carried double the design load in its load test. Yet the structure had cost considerably less to build than a steel or a stone bridge of the same dimensions. He ended his article with a discussion of the three aspects of bridge design that would guide his own work over the next forty years: the empirical proof of efficiency by load test, the social ethic of minimum cost, and the visual·elegance possible in efficient and economical design. Because of the cracking in the Châtellerault Bridge, Maillart would base his own major bridge designs on forms different from those of Hennebique, but the pioneering work of his predecessor had been essential both to show the wide variety of applications possible and to stimulate Maillart to devise his own competing systems.

Fig. 2-6. Vienne River bridge at Châtellerault by Hennebique, 1899

Fig. 3-1a. Veyron Brook bridge

3 The Turn from Masonry (1894-1901)

Pampigny: The First Bridge

As Maillart's studies drew to an end in March of 1894, he found work with the firm of Pümpin and Herzog in Bern.[1] In October of 1895 the firm sent Maillart to Morges, a town just west of Lausanne on Lake Geneva,[2] where they had built a private rail line for a company known as B.A.M. (Bière-Apples-Morges). The section from Morges through Apples to the village of Bière had been opened in June[3] and Pümpin and Herzog were to build a branch line from Apples through Pampigny to the village of L'Isle. Since Maillart had evidently worked in the Bern office on the layout of the line, as well as on the design of its bridges, the firm sent him to the branch site as an assistant to the supervising engineer.

Maillart stayed in Morges until March 1896, when he moved to Pampigny. Here, between March and September he supervised the construction of his first bridge, having worked on the plans in late 1894.[4] Crossing the narrow Veyron Brook, its form, shown in Figures 3-1a and 1b, is entirely traditional and consists of a 6-meter span semicircular arch made of reinforced concrete blocks 60 cm thick.[5]

The new line opened in September[6] after which Maillart returned to Bern. He apparently began to think of other work right away. The firm wrote him a letter of recommendation describing him as a "swift and smooth draughtsman";[7] in January 1899 he took a job with the city of Zurich in the *Tiefbauamt*, the heavy construction division of the department of public works.[8] Maillart was soon given the opportunity to design his first major work, the Stauffacher Bridge over the Sihl River.

Stauffacher: The First Competition

The first design to be considered for the Stauffacher Bridge was for a three-span steel girder bridge, but it was rejected by the cantonal authorities because the two river piers would have obstructed the ice flow. The heavy construction division then made three more designs: a two-span steel arch bridge (cost estimated at 245,000 Swiss francs), a one-span steel arch bridge (268,000 francs), and a two-span masonry arch bridge (272,000 francs).[9]

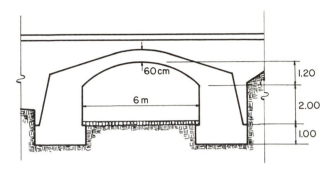

Fig. 3-1b. Bridge of 6-meter span over the Veyron by Maillart, 1896

Still undecided, the city requested in the summer of 1898 that the entire project, especially the steel design, be reviewed by Wilhelm Ritter.[10] Ritter's report summarized the possible options for an urban bridge of moderate size and set out criteria upon which the final choice should be made: "not only usefulness and carrying capacity but also aesthetic considerations."[11] After discussing at length the carrying capacity of the two-span steel bridge design, he gave a long analysis of its aesthetic value, expressing a preference for an odd number of spans because "one opening in the middle is a small-scale symbol of the fundamental object of a bridge: crossing a valley. One pier in the middle looks like an island in the river."[12]

The importance of this Swiss image of crossing a valley led Ritter to argue strongly for a single-span bridge. After rejecting the idea of masonry as too costly for an arch with such a low projected rise, he recommended a three-hinged concrete arch with steel hinges at the crown and at each of the abutments (see Fig. 2-3b).[13] Ritter explained that the calculations for such a bridge would be simple, the load capacity great, and the hinges would permit the arch to adjust to foundation settlements and temperature changes without causing stress. Ritter further recommended a light reinforced concrete roadway of the Hennebique type and noted that a study should be made to determine whether the extra foundation costs for two shore abutments would be less than the savings gained by eliminating the central pier.

On August 13, 1898, the city engineer, V. Wenner, sent a memorandum to the city building commission stating that the heavy construction division would now complete the designs for all three projects: the original two-span steel bridge, the concrete arch with three hinges, and the steel arch, also with three hinges.[14] Wenner clearly followed Ritter's suggestions and Maillart set to work designing his first three-hinged concrete arch bridge. Maillart's design at 220,000 francs was cheaper than any other. According to Wenner, the steel arches would each have cost 240,000 francs and the original two-span design would have cost 265,000 francs, up from the previous 245,000 estimate.[15]

Receiving approval, Maillart designed the concrete arch without reinforcement, following the precedent set several years earlier in Geneva with the Coulouvrenière Bridge, for which Ritter had also been a consultant.[16] For aesthetic reasons, however, the Zurich city architect, Gustav Gull, designed a masonry facade to conceal the concrete structure completely (Fig. 3-2). The arch is basically a solid curved slab of nearly constant thickness, going from 78 cm at the crown to 94 cm at quarterspan and 72 cm at the supports. The light, reinforced concrete deck carries its loads to the vertical cross walls connecting the deck with the arch, and the arch transmits

Fig. 3-2. Stauffacher Bridge over the Sihl River in Zurich by Maillart, 1899

Fig. 3-3. Stauffacher Bridge showing deck, cross walls, arch, and hinges

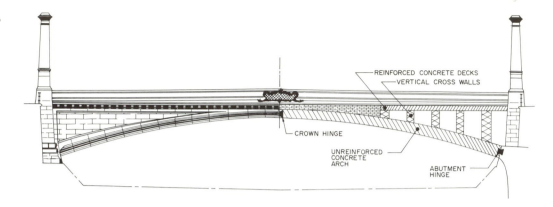

REINFORCED CONCRETE DECKS
VERTICAL CROSS WALLS
CROWN HINGE
UNREINFORCED CONCRETE ARCH
ABUTMENT HINGE

these loads to the abutments (Fig. 3-3). In principle, the structure works like a masonry bridge in that each upper part puts weight on the next lower part. Because there is no reinforcement in the arch, it must carry these loads solely by compression, as if it were made of stone blocks.[17]

This first major structure of Maillart's did not depart significantly from previous designs but the economies were considerable.[18] Wenner was impressed, and in a speech in February 1899, he gave his young assistant the unusual honor of full credit.[19] Yet only Wenner is credited on the official bridge documents and on the figure of the bridge published in a Maillart commemorative booklet by Professor Mirko Roš.[20] Maillart appears to have been prepared for more responsibility, and apparently Wenner, by raising his salary, tried to satisfy him. But Maillart found a new opportunity.

The firm of Froté and Westermann had proposed an alternate reinforced concrete arch design for the Stauffacher Bridge, following the Hennebique system, but its cost of 190,000 francs had not included any masonry facade. In October 1899, Maillart left to join this firm. Accepting his resignation, the Zurich city council recorded their regret at the departure of so able a man.[21] That summer, he had been in charge of redesigning the highway for the Zollikerstrasse and the Neumünsterstrasse on the south side of the city. It seems possible that the success of the Stauffacher Bridge generated in Maillart an interest in bridge design that could not be satisfied by such nonstructural work as roadbuilding. If Froté and Westermann did not win the Stauffacher Bridge contract from the city, they did get its designer; and thus in the autumn of 1899, as Maillart's first major bridge neared completion, he took a job that would allow him to develop fundamentally new ideas of structure.

Zuoz and the New Form

One of Maillart's first assignments with the new firm was to help make the calculations for the Solis bridge, a 42-meter-span masonry arch on the Rhätische Bahn line between Thusis and Tiefencastel (Fig. 3-4). This bridge had two significant features: it was the first in Switzerland to be calculated according to the theory of elasticity, the most advanced method of analysis available; and it was the first long-span arch in that country to be constructed in three concentric rings of stone.[22] In the construction of the Solis bridge the arch was divided into three layers, only the first of which needed to be carried by the wood scaffold. Once that initial layer or ring was complete, it could itself act as a thin arch and carry the remaining two layers. In that way the designer could make the scaffold much lighter, since it was to bear only one-third of the arch weight. This saving could be substantial, because the scaffold usually accounted for a large part of the cost of the entire project. Probably the experience at Solis was to give Maillart the idea of designing the scaffold at Zuoz only for the weight of a thin arch slab.

In each of his two previous positions, Maillart seems to have done a variety of works of which only one stood out: first, the Pampigny arch under Pümpin and Herzog; next, the Stauffacher Bridge under the Zurich *Tiefbauamt*. The outstanding work of his third position was the bridge over the Inn River near the little town of Zuoz, eleven miles downstream from the resort of St. Moritz in the far eastern Canton of the Graubünden.

Studies had begun in early 1899 for the straightening of the Inn River, and in the summer of 1900, the district engineer, E. Ganzoni, prepared a steel truss design for a bridge to be built over the new river bed at a cost of 30,000 francs.[23] Early in August, Froté and Westermann suggested an alternate design for the bridge in reinforced concrete, doing so at the request of A. Schucan, the director of the Rhätische Bahn—a narrow-gauge, privately owned railway then under construction in the Graubünden, for which the firm was already involved in bridge construction.[24]

In August 1900, Froté and Westermann made a formal proposal to the Zuoz town council, including a report on the design, and offered a fixed price of about 31,000 francs. The report, almost certainly written by Maillart, emphasized the competitive price, the same freedom from maintenance as with a stone bridge, and the possibility of using thin sections as in a steel bridge, and noted that it would resemble a stone bridge, making "an elegant impression."[25] This traditional aesthetic idea would go together with a new structural concept, in which "the arched slab, the lon-

Fig. 3-4. Postcard from Maillart, September 1903, showing the Solis bridge

gitudinal walls, and the roadway slab *together* form the arch."[26] The integration of previously separate elements in a bridge was a revolutionary innovation, which Maillart argued would produce a lighter, cheaper, and more elegant structure that would, "bring honor and embellishment to the community."[27]

Negotiations over price took more than a month. On September 18 Schucan wrote to Froté and Westermann that they had been awarded the contract for 26,200 francs, a sum reflecting certain changes in construction.[28] Schucan stipulated, however, that the structure must satisfy their consultant, who, as on the Stauffacher Bridge, was Ritter of Zurich.

Maillart was in Paris when the news of the bridge contract came. The firm had sent him there to negotiate an agreement with M. Bonna, a well-known French pioneer in reinforced concrete and a competitor of Hennebique, presumably to allow Froté and Westermann to use Bonna's system in Switzerland.[29] (While Maillart was in Paris he saw the 1900 Exposition; a landmark in the acceptance of reinforced concrete as an established medium for major construction.)[30] Maillart received his firm's approval of his proposals to Bonna together with the news of the contract.[31]

Construction of the abutments began in the autumn of 1900 but was suspended in November because of the cold weather. As frequently happens in public works, the owner had permitted the builder to begin the foundations even before the structural design had been approved. When the district engineer, Ganzoni, wrote to Ritter in early February 1901 asking when his report would be ready,[32] Ritter's response was unusual. "For about two weeks," he wrote, "I have been busy checking the design for the Inn bridge at Zuoz which you sent me. The problem you have posed is not an easy one; otherwise, I should have finished with it much sooner."[33]

The difficulty arose because there was no precedent for Maillart's idea of building the arch slab, the longitudinal walls, and the roadway slab together to form the arch. Ritter, one of the world's leaders in arch bridge analysis, was confronted by a problem to which there was no available solution. He did not send in his report until May 13; but when he did, he recommended that the design be approved with no further change (Fig. 3-5). Only in July 1901, when construction was well under way, did the community of Zuoz and the builder sign the official contract, which included reference to Ritter's May report.[34] Construction of the bridge ended in early autumn, and in October the load test began.

Typically, Swiss public authorities do not accept bridges until they are load tested, a practice not followed in the United States. Besides assuring structural soundness, these tests have two major benefits: they yield data for the design professionals, and

Fig. 3-5. (Opposite page) Inn River bridge at Zuoz by Maillart, 1901

Fig. 3-6. Stauffacher Bridge, sectional drawing showing the arch shaded

they provide a festive occasion for the local community. Undoubtedly because of the bridge's novelty, the district engineer asked Ritter to design and direct the load test personally. Together, Ritter, the district engineer, the building superintendent, and Maillart (as the representative of the builder) carried out the test program. This consisted of measuring the vertical movements of the bridge: first, under its own dead load after the scaffold was lowered; second, under traffic loads simulated by layers of gravel on the roadway; and third, under heavy truck weights. Ritter reported all movements to be satisfactorily small, and the only difficulty arose from several small cracks that appeared near the middle of the bridge next to the crown hinge. Maillart agreed to minor alterations, and Ritter's report closed with this summary:

> The fundamental ideas upon which the bridge has been designed and built can in certain respects be characterized as new. The bridge, being an innovation, will doubtless find many imitators. Under the circumstances, that not everything was perfect should not surprise us; that the resulting defects are insignificant brings honor to the builder. After all is said and done, the community of Zuoz has gained through the new bridge a commodious, reliable, and relatively maintenance-free facility.[35]

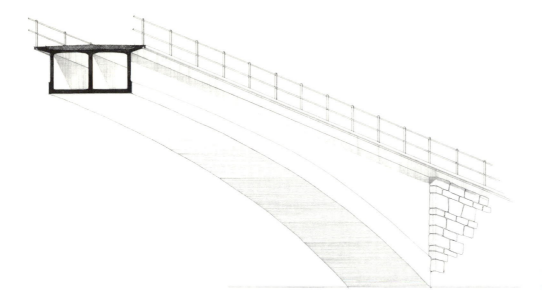

Fig. 3-7. Zuoz bridge, sectional drawing showing the arch as hollow box

The bridge, which gave Ritter so much analytic difficulty, had successfully come through the critical scrutiny of Switzerland's foremost structural authority. But why had it at first puzzled him? Did the bridge have any larger significance? These questions may be answered by referring back to an earlier connection between Ritter and Maillart, the Stauffacher Bridge, and comparing that nineteenth-century Zurich design with the twentieth-century one at Zuoz.

Stauffacher and Zuoz

The cutaway drawings in Figures 3-6 and 3-7 show that in the Stauffacher, the arch alone (the shaded section) carries all the load to the abutments; whereas at Zuoz, the arch slab, the walls, and the deck slab act together (the shaded sections) in carrying the load. In technical terms, the Stauffacher arch is solid, while the arch at Zuoz (made up of arch slab, walls, and deck slab) is a hollow box, the first ever built for a concrete bridge. The Stauffacher deck carries its load to the cross walls, which in turn load the arch, each part acting separately; while at Zuoz, the deck slab and walls act together to carry their own loads and, in addition, to help carry the arch slab. The deck, for example, carries roadway loads transversely to the walls, but in

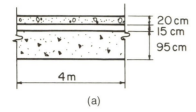

(a)

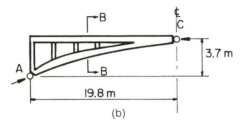

(b)

Figs. 3-8a and 8b. Stauffacher Bridge (a) simplified quarterspan cross section and (b) simplified longitudinal section

addition, helps to carry the same loads longitudinally to the abutments. The deck, therefore, works in two directions at Zuoz, while in only one direction in the Stauffacher. And since the straight deck slab helps the arch slab at Zuoz, the arch slab can be much thinner, varying from 18 to 50 cm; whereas in the Stauffacher, it goes from 72 to 95 cm in thickness. In the Stauffacher, the deck and arch weigh 3 metric tons per square meter of roadway, while at Zuoz, the deck slab, walls, and arch slab together weigh 1.3 metric tons per square meter and the stresses under dead load at Zuoz are lower.[36] In the Stauffacher, the deck adds weight but does not reduce stress; moreover, the masonry side walls merely cover up the concrete structure and add further weight without carrying any load.

For this efficiency at Zuoz, however, Maillart paid a large price in analytical rigor. Any engineer trained in the late nineteenth century could have analyzed the Stauffacher, but the Zuoz bridge presented a formidable problem. In a hollow box, the arch stresses are assumed to be distributed evenly over the cross section. But in practice, although this assumption is correct at the crown (C) and reasonably correct at the quarterspans (B), it is not correct at the abutment hinges (A). (See Figs. 3-8 and 3-9.) In the cross section of the abutments, all the stresses are concentrated at the hinges, with no longitudinal stresses on the deck above or in the longitudinal walls. There was, however, no way in 1901 to determine analytically just what the stresses were between the abutments and the quarterspans. Even today, with electronic computers, it is not an easy problem. Maillart traded analytical rigor for efficiency of design, ease and certainty of mathematical analysis for a more rational use of materials. The Stauffacher Bridge could be justified by a well-accepted theory, whereas the proof of the Zuoz design, as with most of Maillart's later bridges, rested much more on full-scale load tests. Maillart could not convince his doubters with mathematical arguments, because such research was way behind his design practice.

Fortunately, Wilhelm Ritter recognized that good design did not necessarily require rigorous analysis. Without Ritter's support, it is unlikely that Maillart would have had the chance to build his early arch bridges, which were the seeds of his mature works of the 1930's.

The Zuoz bridge was revolutionary in its structurally efficient use of materials. Of equal importance, however, were the new possibilities the bridge opened in economical construction as well. Apart from the obvious differences between a bridge in Zurich, the largest and richest city in Switzerland, and one in the little river town of Zuoz, the economics of construction were very different in the two bridges.

For the Stauffacher, the designer had no direct responsibility for building, and there is no record of any explicit construction method having been prescribed. At Zuoz, however, the designer worked for the builder. As a result, the relationship between design and construction, between theory and practice, was intimate and conducive to cost-saving methods. Maillart's idea here was to build only the bottom curved slab first. Because that slab was so thin, the scaffold could be very light. Once this curved slab had hardened to sufficient strength, the longitudinal walls and roadway deck were cast. In that way, he designed the scaffold to carry only the bottom slab, which in turn carried the rest of the superstructure, with the scaffold serving merely as bracing. Once hardened, the entire structure could then carry roadway loads.[37] This procedure followed from that used for the Solis bridge.

This sequence reduced scaffold costs but introduced major uncertainties into the analysis, because it is very difficult to show mathematically just how the superstructure load is really carried. Whether Maillart was ever forced to produce calculations for this procedure we do not know, but it certainly worked and saved on scaffolding. For most designers working in an office with numbers, such a procedure would seem to complicate design, and usually it would not be considered. Complexity of calculation often has a larger influence on design form than does potential construction simplicity.

Finally, in addition to considerations of efficiency and economy, the two bridges represented a radical difference in aesthetic vision, although not, surprisingly, in appearance. They symbolized two distinct aesthetic attitudes toward reinforced concrete. The Stauffacher embodied the attitude that structure and decoration are separate. The popular desire for elegance was met in that nineteenth-century bridge without reference to the structural material. There was a reflection of the arch form in the flat underside curve, but the stone facing eliminated the texture of concrete, the articulation of the hinges, and the openness of the arch and cross-wall form. Moreover, the four massive abutment pilasters and the heavy stone parapets heightened the sense of cut stone and distracted the eye from the bridge structure. Rather than embellish the existing structure, as was the case in ancient works, extra structure was added for embellishment.[38]

The Zuoz bridge, on the other hand, represented an attempt to create an object in which structure and decoration were fully integrated. The similarity of appearance between the 1899 and 1901 bridges hides the different underlying principles. Had Maillart designed the Zuoz bridge more rationally (from a scientific point of view), it would have looked strikingly different, as his 1905 Tavanasa bridge would. The ex-

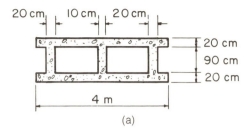

(a)

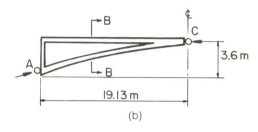

(b)

Figs. 3-9a and 9b. Zuoz bridge (a) simplified quarterspan cross section and (b) simplified longitudinal section

planation of the similarity seems inescapably connected to Maillart's experience with the Stauffacher. After dispensing with its obvious nonstructural pylons and parapets, he simply used the remaining visual elements as structurally as he could to design the Zuoz bridge.

The Zuoz bridge was also a test case to discover how reinforced concrete endures, not as an inert piece in some laboratory, but as a fully exposed structure in a hostile environment. (It is a sad fact that many structures built in the first years of the twentieth century have not lasted well. The Zuoz bridge itself suffered some damage due to weathering, and was rehabilitated in 1968 without, however, altering its form.)[39] In fact, from the perspective of endurance, covering the arch with a solid wall at Stauffacher was rational, while building a solid concrete wall right out to the abutments at Zuoz was irrational. The significance of the Zuoz bridge was not that it was rational or structurally honest or free of extraneous decoration; but that it was the first serious attempt in the history of concrete structures to build a bridge of reinforced concrete by connecting all parts together, both physically and visually.

Zuoz and Isar

To see what a difference this makes, one has only to compare the Zuoz bridge with the three-hinged arch bridge over the Isar River (Fig. 3-10) designed by Emil Mörsch and completed in 1904.[40] Despite the absence of decorative sidewalls, the structural idea was similar to Stauffacher (see Fig. 3-2); the arch alone carried the weights of the deck and columns. Mörsch took pride in the appearance of this bridge, for as he said, "The architecture of the bridge is completely determined by its construction. With the exception of the center one, no pier is at all decorated. All concrete surfaces were left unfinished, except one prominent ridge which was formed by a crack between the form boards."[41]

The difference is not one of applied decoration, but of the actual structure itself. In overall concept, the Mörsch design was a masonry bridge in which the deck carried the roadway loads, the columns carried the deck loads, and the arch carried the column loads. The structure consisted of three parts piled on top of each other, each part well designed out of reinforced concrete, but the whole still conceived along lines not essentially different from the Roman Pont du Gard built nineteen centuries earlier.

The curious paradox is that Maillart's bridge form appears to have been suggested by the decorative solid wall of Stauffacher, but actually represented a radical

Fig. 3-10. Isar River bridge at Grünwald by Mörsch, 1904

change in structural ideas; while Mörsch's bridge form came from the type of structure beneath the decoration of the Stauffacher, but his result represented a thoroughly traditional vision of structure. To turn the paradox around, Maillart was in the
oldest tradition of great building, that which makes decoration out of useful structure;
whereas Mörsch was at the beginning of the twentieth-century tradition of denying
decoration under the misguided notion that elegance will somehow automatically
appear out of the resulting bare structure.

This latter idea, which became something of an ideology for early twentieth-
century artists and architects, found its purest expression in a 1908 article by the
Viennese architect Adolf Loos, entitled "Ornament and Crime." "I have . . . evolved
the following maxim, and pronounce it to the world: the evolution of Culture marches

Fig. 3-11. Maria Ronconi-Del Santo (1872-1916) at the time of her engagement to Maillart, 1901

with the elimination of ornament from useful objects," Loos wrote. This statement[42] signaled the growth of an idea about design that had its roots in attitudes about engineers. As Rayner Banham observes:

> Freedom from ornament is the symbol of an uncorrupted mind, a mind which he [Loos] only attributes to peasants and engineers. In this view succeeding generations were to follow him, thus laying further foundations to the idea of engineers as noble savages and also—and this is vital to the creation of the International Style—laying further foundations to the idea that to build without decoration is to build like an engineer, and thus in a manner proper to a Machine Age.[43]

Like Loos, Mörsch evidently believed that by removing all decoration from a technically correct structure, a new form would result. However, the imaginative creation of structural form requires a positive principle, not merely a negative stripping down. Great designs are not the product of negative ideas, such as the absence of decoration, or a minimum of labor and materials. Decoration, properly understood, begins with the desire to choose from among many possibilities the most pleasing form. For Maillart decoration meant the expression of efficient, elegant forms.

There is no way to interpret Maillart's designs, especially his later ones, without recognizing his strong expressive urge. No automatic single solution led scientifically to his forms. These were suggested by what he saw in traditional building and by what he felt would make a stunning image. His imagination selected a form from the set of scientifically reasonable possibilities which, although limited, could not be reduced to a fixed number. Maillart thus began a search for new creative possibilities with form, rather than merely the chance to display bare structure.

Between August 1898, when he designed the Stauffacher Bridge, and August 1900, when the Zuoz design was submitted, Maillart achieved a fundamental shift of vision, from which all of his mature works developed. It is a paradox that this new vision seems to have grown out of a respect for the visual elegance of the Stauffacher. In an illustrated lecture delivered at Basel in 1938, he began with two slides of the Stauffacher for which his notes say, "Beautiful medieval bridge giving no visual reference of the structure." Then, after showing his first bridge design for Froté and Westermann, the Zurich Hadlaub bridge (1900), he showed the Zuoz bridge, introduced by the comment, "based on ideas from the Stauffacher Bridge."[44]

The Zuoz bridge provided Maillart not only with the opportunity to develop his first major design, but also with another kind of opportunity that changed the course of his life. During the summer of 1901, while the bridge was under construction, he traveled from Zurich to Zuoz to spend time at the site. In his hotel dining room he met a lovely young girl named Maria Ronconi, from Bologna; her parents had died and she was traveling with a Swiss family.[45] Looking at her picture (Fig. 3-11), it is not difficult to see why Maillart was attracted. He seems to have had few previous attachments. He was already twenty-nine, well into his professional career, apparently deep in his work. He was known to be quiet, serious, and intense (Fig. 3-12).

Maria sent him a photograph inscribed "In memory of the days at Zuoz" when she finally returned to Bologna and he to Zurich. A lively correspondence followed, and in late August they were engaged. In Zurich, an announcement in the *Zürcher Wochen-Chronik* of September 7, 1901, indicates that Maillart was beginning to be recognized: "Engineer Robert Maillart from Bern, the gifted builder of our Stauffacher Bridge, has just become engaged to Miss Maria Ronconi of Bologna. Our congratulations!"[46]

On November 11, 1901, the young couple were married in Bern and soon thereafter set up housekeeping in Zurich. Maillart left Froté and Westermann and early in 1902 founded his own firm. With these changes, he rapidly became a successful entrepreneur as well as a designer-builder of unusual structures.

Fig. 3-12. Robert Maillart (1872-1940), c. 1901

Fig. 4-1. Thur River bridge at Billwil by Maillart, 1904

4 The New Bridge Form (1901-1904)

Zuoz and Billwil

The new firm's first contract must have been welcome but was hardly a vehicle for new visual forms: two gas tanks for the town of St. Gallen. The second, also in St. Gallen, was more significant. A bridge over the Steinach Brook, it was in a line of development leading from Maillart's analytical work on the Solis bridge to the Lorraine project of 1911 and the completed Lorraine Bridge of 1930—all of them more important as construction projects than as original designs. The Steinach bridge, unlike the Solis, was built entirely of concrete blocks; even the facing blocks were concrete, with broken natural stone surfaces cast in to give a masonry-like facade.[1]

Of the seventy-four works built by Maillart between 1902 and 1913, only two played a direct role in his later development of a new bridge form using three-hinged arches: the Thur River bridge at Billwil, begun in 1903 and opened in 1904; and the Rhine River bridge at Tavanasa, begun in 1904 and completed in 1905.[2]

The Billwil bridge differs from Zuoz primarily in that it has two spans (Fig. 4-1); otherwise, each of its spans is almost identical to the single span at Zuoz.[3] In the case of Billwil, however, sufficient data exist in the *SBZ* and the archives of the Canton of St. Gallen to reconstruct in detail Maillart's design method.[4] Just as at Zuoz, Maillart designed the full hollow-box section (i.e., horizontal roadway deck, longitudinal walls, and curved arch slab) to carry the load near the crown, with the lower slab to carry it at the springing lines. This made the deep longitudinal walls near the abutments structurally unnecessary, but Maillart, in his computations, did not recognize this fact. Therefore, in May 1903, nearly three years after the completion of the Zuoz design, he had made no basic change in his design ideas; he simply adapted an existing and tested design to the longer crossing at Billwil.

One of Maillart's first actions after establishing his firm had been to obtain a patent on his new bridge system, which he called "Arched Beam of Reinforced Concrete." The patent was applied for in February 1902, the claim stating that the "arched girder of reinforced concrete is characterized by an arch slab, a roadway slab, and longitudinal walls. The walls support the roadway slab and, with the help of steel bars in the concrete, connect the two slabs to form a stiff integral structure."[5] This claim is simply a paraphrase of Maillart's Zuoz design of 1900. His first opportunity to use the patent came with the Billwil bridge.

Billwil barely qualified as a settlement in 1902; there were just five houses with twenty residents.[6] For this reason, when the building authority of the Canton of St. Gallen studied the bridge question, one proposal was for a footbridge only, to replace a defective footbridge built thirty years before over an old ferryboat crossing.[7] The authority decided to build a road bridge, however, and selected a steel design with a central stone pier, which it estimated in January 1898 would cost 54,800 francs.[8] When competitive bidding on the project was announced in 1903, builders were also allowed to submit prices for alternative designs.

In May, Maillart submitted a bid of 39,440 francs for his own plan for the bridge and its approaches, to be built of reinforced concrete. He enclosed a letter describing its superiority over steel because of lower maintenance costs and better appearance, writing that while the budget would permit no decoration on the bridge, the community would still obtain a significantly more beautiful structure than if a steel bridge were to be built.[9] As evidence of the appearance, Maillart included with the letter photographs of the Zuoz bridge, which he compared to his design proposal for Billwil. He observed that conditions at Zuoz, such as soil type, span length, and climate, were worse than at Billwil, but, "in spite of all these unfavorable circumstances [at Zuoz], Professor Ritter and also Director Schucan and Chief Engineer Gilli of the Rhätische Bahn all recommended our system [for Zuoz] even though a steel bridge would have been cheaper."[10]

The canton sent Maillart's plans to Ritter immediately, and on June 17 he responded with a preliminary report recommending Maillart's system over a rival's two-hinged arch design, but urging that the canton write to Zuoz to find out how that bridge was holding up.[11] At the end of June 1903, the building authority of St. Gallen awarded the contract to Maillart. Maillart's arch design appealed to the cantonal engineer, partly because the Thur River had cut a narrow channel there in strong rock which allowed for reasonably short spans on good foundations, and partly because the maintenance of a concrete bridge would be less than for a steel bridge, which would have cost about the same initially but which would need repainting periodically to prevent rust.[12]

In August, Maillart wrote to the cantonal engineer of a change in the exterior facade of the central pillar and the abutments, made because "we have consulted about that with an architect."[13] Although there is no direct evidence, it would appear from this reference that Maillart was forced by one of the authorities to cover the exposed concrete at these places with an ornamental facade (Fig. 4-2). The architect in question was probably Otto Pfleghard of the Zurich firm of Pfleghard and Haefeli, with whom Maillart would collaborate on a proposal for a bridge over the Sihl River in 1904.[14]

Included with this letter were the structural calculations and five drawings, all of which clearly show Maillart's method of analysis, and from which one can see that he was counting on the full participation of the vertical walls, even near the abutments.[15] He made only one significant structural change from the Zuoz design; he increased the amount of concrete at the crown hinge to reduce the stress there to less than half of its value at Zuoz.[16] This was, without a doubt, to avoid the small cracks found by Ritter in the crown at Zuoz during the load test.[17]

Ritter's final, more detailed report on Billwil was devoted largely to the structural calculations, with which he was generally in agreement. He mentioned but found no objection to stresses in the section near the abutments. However, he was concerned about the hinges themselves, again reflecting his observation of cracks at Zuoz. He concluded by remarking that the best guide for constructing this bridge well would be the experience gained at Zuoz.[18]

Maillart responded promptly showing how much more conservative the Billwil hinge design was than the one at Zuoz. He argued that the small cracks had arisen because of the foundations giving way slightly; and he predicted that this would not occur at Billwil because the abutments were on solid rock. Nevertheless, Maillart agreed to place some reinforcing bars vertically in the concrete near the crown hinge, in order to remove the danger of horizontal cracks.[19]

Construction of the Billwil bridge, which had begun in July 1903, was essentially completed by December. The scaffolding, left in place over the winter, was removed and a full-scale load test was carried out in April 1904. A newspaper account described how the local fire company loaded the bridge with 1,520 hundred weights of water, equivalent to the weight of 1,000 men or 76 metric tons (76,000 killograms to make 300 kilograms per square meter of surface area). This corresponded to the design live load for snow, and resulted in very small measured vertical deflections and no cracking. Maillart's predictions in response to Ritter's second report were thus confirmed, and the cantonal authorities accepted the bridge. The reporter praised all concerned. "The firm of Maillard [*sic*] and Cie must definitely be congratulated on the beautiful and complete success it has achieved. Also, we must congratulate the local authorities and the entire community, which for the price of 40,000 Fr. has got such a beautiful, pleasing, and sound public works structure."[20]

Fig. 4-2. Detail of the midstream pier for the Billwil bridge

This short article in a provincial newspaper reflects a distinctive Swiss attitude toward public works, a mixture of civic pride, aesthetic consciousness, and thrift. This spirit, and the interest and encouragement shown by his wife in his projects,[21] helped Maillart make his great contributions to structural design, though his vision was not always accepted even in his own country.

Zuoz and Tavanasa

At Schucan's request, Maillart returned in September 1903 to inspect cracks that had appeared in the vertical walls of the Zuoz bridge, with which the local residents nonetheless remained very pleased, as he found on his arrival.[22]

In his report to Schucan,[23] Maillart recorded the existence of three vertical cracks near the left bank and one large vertical crack near the right bank, both on the sunny side of the bridge. Also, one horizontal crack appeared near each bank, again on the sunny side. As he observed, and as his Billwil calculations showed, the stresses at these points from dead weight and live loads were very small. Maillart concluded that the cracks must have resulted from a tendency of certain parts of the bridge to contract or expand as the concrete hardened over the years. This action, as Maillart explained, occurred when the dry, sunbaked vertical sidewalls contracted inward while the predominantly horizontal arch, continuously wetted by the river, was contracting much less. The deck, wetted by weather, also moved inward much less. (As shown in Figure 4-3, the arch and deck restrained the walls from contracting, pulling the walls in two different directions. The cracks resulted.)

Maillart went on to emphasize that the cracks did not impair the utility of the bridge, and that they might grow for another year and then stop. Maillart suggested that a simple and cheap way to restore the appearance of the bridge was to whitewash the sidewalls.

Thus, while the Billwil construction was under way, Maillart discovered a defect in his earlier Zuoz bridge. There was still time to change the Billwil design if the cracks at Zuoz seemed significant, but instead Maillart stuck to his design submitted several weeks earlier for a bridge with the same solid vertical walls as the Zuoz. There were probably three reasons for this decision. First, neither Maillart nor anyone else considered the cracks at Zuoz to be structurally serious. Second, after having pointed to the Zuoz bridge as a good example, it would have been awkward to have publicized defects in it if they were not critical. Finally, in any case the solution to the

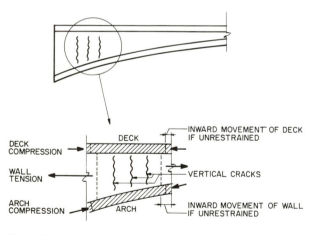

Fig. 4-3. Vertical cracks in the Zuoz bridge

cracking did not yet exist, although Maillart would develop it in his next bridge at Tavanasa.

The physical problems at Zuoz were relatively easily disposed of but a legal problem arose which Maillart could not overcome. His bridge patent was challenged by Froté and Westermann on the grounds that he had developed the idea while in their employ. In late September 1903 it went to court and the judgment went against Maillart, the patent being declared void.[24] Thus, his use of the term *Système Maillart* for arched-girder bridges occured only for Billwil and never appeared in later brochures. His loss in court did not, however, prevent him from using the design idea again.

Not all of Maillart's early efforts to win contracts were successful. Early in 1904, the city of Zurich announced a competition for a new Uto bridge over the Sihl River about one mile upstream from the Stauffacher.[25] This presented Maillart with the chance to build a city bridge using the design system developed at Zuoz, and he entered a design in collaboration with the Zurich architects Pfleghard and Haefeli. They shared with another entry the second prize of 900 francs; there was no first prize. Maillart's design had almost the same dimensions as the Stauffacher and the Zuoz bridges, with a span of 38.4 meters and a rise of 3.3 meters. The longitudinal walls were still solid out to the abutments, but column-like vertical ridges protruded from the walls, probably decoration suggested by the architects. Structurally, the proposed bridge was identical to Billwil, except that the longitudinal exterior walls were 50 percent thicker than the interior walls, a change that may have been Maillart's first answer to the Zuoz problem of cracking. But the other second prize, a two-span, stone-covered, hingeless concrete arch, was eventually chosen for construction, and indeed Maillart was never to build an exposed-concrete bridge in a major Swiss city. The various competition designs went on public display in June, and Maillart wrote to Maria, "Our design does not at all please the general public, who insist upon the banal and the ordinary."[26] By September, having lost the Uto bridge competition, Maillart wrote his wife that he had no bridges and was, therefore, forced to send some photos of a factory he was building.[27]

Earlier that summer, he had visited the little town of Tavanasa, from where he sent home a card with a picture of an old wooden bridge crossing the Rhine.[28] When a competition for the Tavanasa bridge was announced in the autumn, he began the design for a new concrete arch bridge at this crossing; and by then his ideas about the Zuoz cracks had changed significantly.

Fig. 4-4. Rhine River bridge at Tavanasa by Maillart, 1905

Tavanasa

Billwil was Maillart's only three-hinged arch bridge of more than one span. He had divided its 70-meter opening into two, necessitating a midstream pier, probably because the resulting 35-meter spans were so close to the 39.6-meter span at Stauffacher and the 38.5-meter span at Zuoz.[29] The total opening at Tavanasa was 51 meters, which forced Maillart to choose between two very much shorter spans of about 25 meters each, or one span almost 30 percent longer than Stauffacher. There was a precedent set by Hennebique's Châtellerault bridge of 1899 for a span with such a low rise. In fact, Hennebique's bridge had a span-rise ratio of 10.4 compared to 8.4 for Billwil and 9.0 for Tavanasa. Thus, the scale and overall dimensions of Tavanasa are not remarkable. The structural system was the same as Billwil, with the deck, walls, and arch integrated.

The difference was mainly that at Tavanasa Maillart cut out from the longitudinal side walls just those regions in which the cracks at Zuoz had appeared (Fig. 4-4), thus removing material that at Zuoz had carried little load and had created an un-

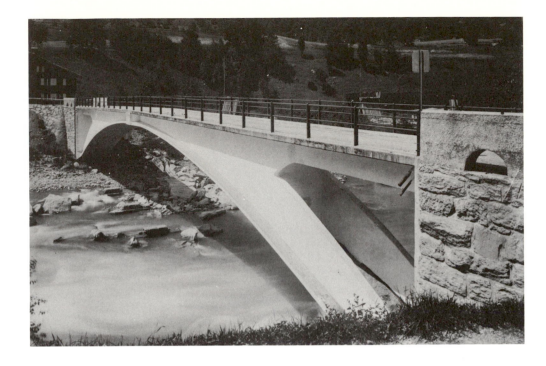

Fig. 4-5. Rhine River bridge at Tavanasa by Maillart, 1905

pleasant visual problem. At Billwil, some load went through the walls, but not much in the region near the edges. If this sounds imprecise, it is. Perhaps this is the reason for the omission of a force diagram from the *SBZ* article on Billwil, unlike the corresponding article on Tavanasa.[30] Maillart removed one ambiguity by cutting out part of the wall and showing, abstractly in the force diagram, and concretely in the structure, that the wall is really part of an arch. But this new choice of concrete abutment form introduced further ambiguity. The horizontal deck and ribs that remain above the cut-out seem to be built together with the abutment wall; but actually they merely rest on the top of the thin cross wall (visible at the left abutment in Fig. 4-4), which in turn is separate from the stone abutment. Visually, the bridge appears to be supported at the abutments by both the horizontal deck girder and the curved tapering arch. This interpretation is possible even when the walls are solid, for as Mörsch had written in 1908, "According to a method of construction already much used in Switzerland by Maillart of Zurich for reinforced concrete arch bridges, the side walls and the deck, which both consist of reinforced concrete, can be built in cantilever form decreasing in [vertical] depth from the abutments to the crown."[31]

This visual ambiguity between cantilever and arch arises largely because of the solid heavy stone abutments (Fig. 4-5), a feature of all four bridges from Stauffacher to Tavanasa, into which the walls of the first three and the top structure of Tavanasa appear to go for horizontal as well as vertical support.

The deck was designed primarily to withstand the live loading of the heaviest single vehicle which, in the case of Tavanasa, was a truck weighing 6 metric tons (13,200 pounds). The concentrated wheel loads had to be carried by the thin deck to the deep walls in the transverse direction (i.e., perpendicular to the direction of traffic). Because the roadway was narrower at Tavanasa than at Billwil (3.2 meters instead of 3.6 meters), Maillart had decided to eliminate the central longitudinal wall and increase the deck span in the transverse direction from 1.26 meters to 1.96 meters (Fig. 4-6). This made it necessary to increase the deck thickness from 10 cm to 12 cm.[32]

Fig. 4-6. Tavanasa bridge, sectional drawing

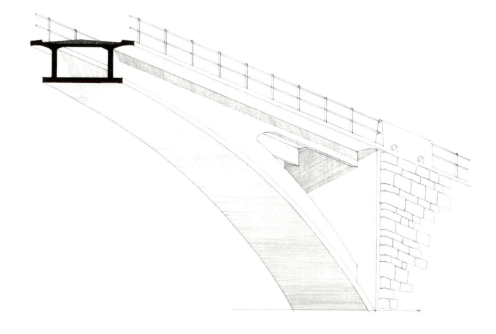

In addition to providing support for the deck, the walls served as part of the overall arch. Whereas the deck carried essentially truck loads only, the overall arch carried essentially dead loads only. Therefore, the reduced walls meant reduced dead loads and hence reduced stresses. The smaller the cross-sectional area, the greater the stress; and the smaller the forces, the smaller the stress. When these two effects are combined, they cancel out and hence the idea that for dead weight in arches of the same span and rise, the stress in the arch material is independent of the cross-sectional area, which means that the area should be made as small as it is possible to build.[33] According to Maillart's force diagram, the total force at the crown under dead load at Tavanasa was 237.8 metric tons, which resulted in a stress of 20.9 kg/cm². Comparable values at Billwil were about 159 metric tons and 12.7 kg/cm², which stresses are just about exactly in the proper ratio relative to the difference in span and rise.[34] Within both bridges, Maillart tried to keep the stresses roughly the same everywhere, and since the forces increase as the arch curves down from crown to springing lines, that is, at the abutments, the section would also need to increase for dead loads.[35]

The overall arch section had, in addition to its dead-load requirement for an increasing cross-sectional area, a separate requirement for live load. Published documents give no indication that Maillart considered the live loads in the derivation of his structural forms for any of his bridges built between 1899 and 1913. Nevertheless, it would be possible to explain the origin of the Tavanasa form by reference to live loads instead of dead loads. Even though this seems not to be the explanation for Maillart's form, discussion of live loading shows that the Tavanasa form did have the additional advantage of a deeper quarterspan section where live-load bending is greatest. Maillart was certainly aware of this advantage at least as early as 1898 during the designing of the Stauffacher Bridge, for which he made the quarterspan arch section the largest.

In nonscientific terms the dead load, being merely a matter of calculating known weights, permits the designer to avoid completely any dangerous tension stresses by making forms that carry themselves entirely by compression. This dead-load criterion of pure compression explains the basis for form throughout the history of masonry structures; that is, those structures built of cut stones across whose joints no tension or pull can exist.[36]

The live load, on the other hand, is accurately known neither in size nor in distribution. For example, whereas the dead load is symmetrical about the midspan of Maillart's arch bridges, the live load can easily appear on only one half of the span.

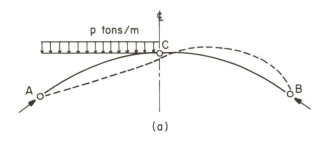

(a)

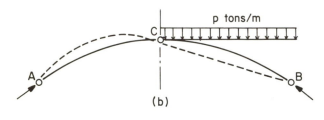

(b)

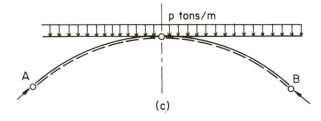

(c)

Fig. 4-7. Deformations of three-hinged arches

And whereas the size of dead load depends only upon the dimensions and the densities of the bridge materials, the live load magnitude, although specified in legal codes, can vary widely and can never be completely controlled. The designer needs to take this difference into account. Where the bridge dead loads were from heavy solid stone, and the live loads were from people, horse carts, and snow, the dead load dominated form making. With the introduction of light metal and then thin reinforced concrete for dead load, along with railroad trains and horseless trucks for live load, the importance of live load increased, particularly for relatively short-span bridges.

For three-hinged arches, the live load is most critical when concentrated all on one side of the span; then the loaded side bends downward and the unloaded side upwards. If the first load is uniformly distributed (See Fig. 4-7a) over half the span, then the loaded half bends downward and risks tension stresses on its underside, whereas the unloaded half bends upward risking tension on its upperside. When an equal uniform load is added to the right-hand side, the influence is reversed (Fig. 4-7b) so that the combined result yields no bending at all and the arch carries the load entirely by axial forces of pure compression (Fig. 4-7c).[37] Figure 4-7a corresponds to a potential live load—the bridge half loaded by vehicles or snow—whereas Figure 4-7c represents a typical dead-load case.

Furthermore, the bending stresses will be largest about midway between the hinges, that is, at the quarterspans, and they will be zero at the springings and crown because the hinges allow free rotation without any stresses due to bending. Therefore, to reduce live-load bending stresses, the designer needs to increase the arch section towards the quarterspans, which is exactly what Maillart did at Stauffacher where the quarterspan thickness was 95 cm while the crown and springing thickness were 78 cm and 72 cm respectively. In just the same way, Mörsch, in his Isar bridge, designed the quarterspan solid arch to be 1.20 meters thick with the crown and springings 75 cm and 90 cm respectively.[38] The analytic principles for this design idea appeared at least as early as 1908 in an article by Josef Melan, who described the best form for a three-hinged reinforced concrete arch as one having a small crown and springing depth, but a substantially greater depth at the quarterspans.[39]

Returning to Tavanasa, we can now see its profile as growing out of this live-load criterion whereby the section depth is minimum at the crown and springing lines and maximum at the quarterspans. What the evidence shows is that however clearly Maillart recognized this live-load advantage, he developed the form from a dead-load criterion. He simply eliminated material in the longitudinal walls near the abutments, where cracks had arisen because of long-term changes in stresses (due to the tendency of different parts of the structure to deform differently).

Later, when Maillart returned to the question of bridge design after a hiatus of nearly twenty years, he would take up the question of the live-load criterion, and as with the dead-load criterion between 1898 and 1904, he would create a completely new form.[40] To that story we shall return later; for now it is only important to emphasize that the Tavanasa form arose out of his three previous bridges. Tables 4-1 and 4-2 summarize these developments and give dimensions and costs, along with similar data for two major pre-1904 bridges, one by Hennebique and one by Mörsch.

Table 4-1
Dimensions of Maillart's Three-Hinged Arch Bridges: 1899-1905

	Span (in meters)	Rise (in meters)	Span/Rise
Stauffacher, 1899	39.6	3.7	10.7
Zuoz, 1901	38.25	3.6	10.6
Billwil, 1904	35	4.2	8.4
Uto (not built), 1904	38.4	3.3	11.6
Tavanasa, 1905	51.25	5.7	9.0

Table 4-2
Costs of Some Concrete Arch Bridges: 1899-1905

	Total cost	Roadway Surface Area (in meters)	Cost per Meter Squared	Dollars per[a] Foot Squared
Stauffacher	220,000 Sf	20 × 51.3	214 Sf	$3.96
Zuoz	27,000 Sf	4.6 × 47.68	123 Sf	$2.28
Billwil	40,000 Sf	4.04 × 84	118 Sf	$2.19
Uto (not built)	145,500 Sf	16 × 54	168 Sf	$3.12
Tavanasa	28,000 Sf	3.6 × 61	128 Sf	$2.36
Châtellerault	175,000 Ff	8 × 130	168 Ff	$3.12
Isar	260,000 Dm	8 × 230	141 Dm	$3.27

[a] Conversion from European money into dollars follows Baedeker's 1905 ed. of *Switzerland* where $1.00 = 5 French francs = 5 Swiss francs = 4 German marks.

The competition for the Tavanasa closed on October 1, 1904 and Maillart won a lump-sum contract on the basis of his own design.[41] Construction began in the spring of 1905; June saw the arch and side wall forms in place. In August a storm nearly destroyed the scaffold, but by September all was complete and the load test was successfully carried out.[42] Figures 4-8 to 4-10 show stages in the construction of this bridge, the last one Maillart was able both to design and build following his own ideas of structural form.

Fig. 4-8. Scaffold and arch forming for the Tavanasa bridge, May 1905

Fig. 4-9. Forming for longitudinal walls for the Tavanasa bridge, June 1905

Fig. 4-10. Completely cast structure for the Tavanasa bridge, late summer 1905

5 From Bridges to Buildings (1904-1909)

The Billwil and Tavanasa bridges revealed two sides of Maillart's practice: the business objective of building competitive public works, and the creative desire to pioneer the artistic use of reinforced concrete. The problem was not uncommon: Einstein was forced to divide his monumentally creative early years between his job as a patent clerk and his research as a theoretical physicist, and Paul Klee worked as a teacher of architects to support himself as a painter.

When Maillart began his own firm in 1902, he must have realized that bridge-building alone could not keep him in business. Only eleven of the seventy-four works attributed to him between 1902 and 1913 were bridges.[1] The bulk of his practice consisted of other kinds of reinforced concrete structures. Bridges could not have sustained him partly because Switzerland needed too few for the relatively large number of engineers and partly because Swiss bridges are mostly small, giving small profits.

We have seen that the use of reinforced concrete was rapidly increasing at the turn of the century. By 1899, only nine years after Wayss began constructing reinforced concrete arch bridges, his corporation had completed about 400 Monier arch bridges.[2] By 1902, ten years after beginning his own firm, Hennebique was operating a huge international network with more than 1,500 contracts per year.[3] It was a time of high activity, new ideas and a widening range of applications; code committees were formed and textbooks written, putting the results of research and practice directly into the hands of thousands of interested persons.[4]

When Maillart began his firm in 1902, it was an exciting time to be working in reinforced concrete, with many opportunities and very little restriction on either the scope of application or the methods of design. This prospect appealed to him, both for its design possibilities and for its business potential. After two years, with Billwil completed and Tavanasa designed, Maillart was established in Switzerland as a leader in reinforced concrete structures, and was invited by the Swiss Society of Cement, Lime, and Gypsum Manufacturers to give the only general address devoted to concrete structures at their 1904 meeting. In September, Maillart gave his first major statement on reinforced concrete at this meeting in Basel. His emphasis on nonbridge structures showed how his career was changing.[5]

He began his speech by saying that the building in which the meeting was taking place had stimulated a negative attitude towards reinforced concrete. The Hotel Bären was such a bad example of recent concrete construction that some would have expected him "to present a funeral oration to reinforced concrete."[6] But he quickly described the reasons for a positive approach, after warning that not only economy, but good workmanship was essential to the success of any new building material.

Maillart said that the use of reinforced concrete for projects requiring fireproofing and large carrying capacity was already established; possible new applications included gas and water tanks. He could illustrate this from experience, for, apart from the two bridges, all his firm's works prior to 1904 had been such tanks; and of the thirteen projects his company had completed by the end of 1904, seven were tanks.[7] Maillart also discussed the importance of reinforced concrete for pipes, largely because of the ease of transportation and installation. Then followed the most significant paragraph of all:

> Well known to all of you is the application of reinforced concrete to bridge construction: small girder bridges are already to be seen here and there, even for railroads. Arch bridges are less frequently to be seen, in spite of the very successful example of a Monier bridge which has already been standing for a dozen years near Wildegg. In the last several years, about three bridges have been newly built in reinforced concrete with spans of from thirty to forty meters. This type of structure is represented by the important Chauderon bridge in Lausanne currently under construction.[8]

In this offhand way, Maillart disposed of arch bridge construction, presumably including his own three bridges, Stauffacher, Zuoz, and Billwil, without mentioning his new bridge at Tavanasa, which must at least have been in design by then. He singled out the older Wildegg bridge and the new Chauderon bridge, the former in a system he had not followed, and the other in a form of only minor consequence to the development of twentieth-century bridge design.[9]

Maillart's main interest at that time was not in the development of bridge design nor in expounding his own original ideas. Instead, he wanted to extend the application of reinforced concrete to other structures, which he illustrated with examples mostly outside Switzerland, and which had greater potential as business propositions than as aesthetic design forms.[10] These advances included concrete pipe and pipe under internal pressure, factory-made pipe sections, high concrete chimneys, concrete masts for high tension power lines, concrete piles for foundations, and concrete railroad ties. As the numerous examples in Paul Christophe's 1902 book *Le Béton armé* show, and as Maillart himself admitted, none of these examples was new.[11] But in Switzerland, they were still not well known in 1904. Maillart then contrasted concrete pipe with cast-iron pipe, concrete chimneys with brick chimneys, and concrete masts with wooden poles and metal latticework, all to illustrate the superiority of reinforced concrete as a building material for a wide range of uses. His only remarks about structural appearance concerned concrete masts: "slender obelisks [that] make a peaceful and pleasant impression," unlike the "disquieting" latticework masts.[12] Maillart's firm at that time was already working on about 150 such concrete masts for the city of Zurich, and would do several similar projects later.

Maillart's business preoccupation with nonbridge structures from 1902 to 1913 was further described in other public statements (see Appendix A).[13] An article in the *SBZ* presented in detail a commercial application (a buried pipe) in which the structural form was technically important but visually insignificant. In 1907, the city of Basel, while building a new railroad station, had to build an underground conduit 100 meters long in order to take a brook beneath the station and the adjacent street (Fig. 5-1). The department of water works accepted Maillart's proposal of a conduit with an elliptical cross section, provided he would test to destruction a three-meter-long segment to confirm the design. As Maillart wrote, the test was necessary because a satisfactory analysis was not practicable.[14]

The test proved the adequacy of Maillart's design and allowed him to draw three conclusions: the specially curved profile was suitable; the flexibility of the soil played a crucial role in the way the arch form deflected under the load; and a thin section was preferable to a thicker one because the latter, being stiffer, would crack more under soil settlement. The ideas connected with this test would continue to guide Maillart, and can be summarized in three principles. First, structural strength is derived from form rather than from materials. Second, field and test experience take priority over theoretical and mathematical analysis. Third, maximum quality goes together with minimum materials.

Figure 5-2 shows the test section after collapse and it clearly indicates how the top arch broke at three places, the crown and the two springing lines, thus forming in effect a three-hinged arch at failure. Although it would not be practical to insert

Fig. 5-1. Riehenteich conduit at Basel by Maillart, 1907

Fig. 5-2. Test section for the Riehenteich conduit

hinges in long pipe, this photograph demonstrates the idea of hinging that Maillart had developed for his bridges.

Some of the ideas implied in his 1907 *SBZ* article became more explicit in Maillart's first speech before the Zurich Society of Engineers and Architects in 1909, as reported in the *SBZ*:

1. Theoretical methods used for other materials are worthless for calculating the safety of reinforced concrete structures. These methods have practical value only if by stresses [the results of the mathematical calculations] are meant [approximate] numbers for guidance rather than actual facts.
2. A calculation method for flat-plate structures [such as floor slabs] has still to be created through tests, in order to make possible a rational use of materials.
3. In the calculations for reinforced concrete structures, in view of the variety of external forces, there are so many different possibilities that it is not sensible to

establish strict [building] codes. On the contrary, a certain freedom must be maintained for the designer. To avoid misuse, the design of reinforced concrete structures should be left only in the hands of the experienced and the conscientious.

4. The designers should be urged to sharpen their practical experience through load tests in which they establish the deflections of the structure, in order to learn to what extent the initial design assumptions for their calculations prove to be correct.

5. The safety of reinforced concrete construction can be satisfactorily guaranteed by modern methods and experience, and criticisms of it have been shown invalid.[15]

These five points are neither independent principles nor of equal importance, but in them Maillart expressed for the first time a coherent set of ideas about structural design: theory is dangerous, numbers are merely guides, codes are restrictive, full-scale testing is crucial, and safety can be guaranteed. His basic idea was that reinforced concrete is so unpredictable that only from direct observation of the material in action can good designs result.

From this fundamental principle, Maillart had developed one special application, the flat-plate structure; and he had done it essentially without theory, without using stresses as a control, without any codes, and on the basis of full-scale tests. The above speech, in effect, articulated the ideas from which Maillart had developed his most successful system to date: the flat-plate "mushroom" slab, which greatly improved his business in pre-World War I Switzerland, and which would help sustain him in his vastly different life after 1920.

Before turning to the "mushroom" slab in the next chapter, it is necessary to understand more fully why Maillart spoke so strongly about the dangers of theories and codes. By 1909, both had achieved a currency that would last for half a century. Both, by their misuse, were evasions of the realities of building. Theory involves both generality and abstraction; theory in engineering tries to reduce the study of specific problems to the solution of general equations. The danger in this arises when the designer's energy goes into the equations rather than into the actual construction. The designer risks becoming a mere analyst, trying to derive practical forms from abstract mathematical formulations. The other side of this problem is codification. Codes are devised to insure legality and safety; the danger is that they reduce specific problems to a choice of standardized forms. The designer relying too heavily on codes risks becoming a mere copyist, trying to use forms that fit the codified formulas.

Paul Christophe provides, perhaps, the most dramatic example of the misunderstandings that arise when theory takes priority over experiment. In his 1902 book *Le Béton armé*, he presented the Hennebique methods for designing columns, and then, on the basis of theory, criticized strongly what his former employer had done and gave another method of his own.[16] Christophe's method achieved general acceptance well before 1909, and its theory entered into existing codes. For example, in America all major codes and textbooks accepted the method given by Christophe, and almost all concrete structures built until Maillart's death in 1940 are supported by columns following this approach.

Gradually, however, laboratory tests began to indicate that something was wrong with this theory.[17] In 1940, the American joint code returned to Hennebique's original method; but, in more general terms, Hennebique's approach to concrete structures, based upon his own tests from 1880 to 1892, was made the single basis for the American code only in 1971. Thus, the influence of theory was so powerful that it took nearly three-quarters of a century to adopt the approach for which Maillart was arguing in 1909.

6 The Business of Building (1909-1919)

Maillart did not neglect bridge contracts entirely. In 1907 he built two small, beam bridges over a railway at Aach, in 1909 he constructed a three-hinged arch at Wattwil with architectural decoration designed by the Bern firm of Joss and Klauser, and in association with the same firm he built two bridges of concrete blocks over the Rhine, one at Laufenburg in 1911 (Fig. 6-1) and the other at Rheinfelden in 1912 (Fig. 6-2). For both of these his competition entry had received only second prize, but his economical method of concrete-block construction derived from the Steinach bridge had probably carried the day. Nearby he had built a bridge over a canal at Wyhlen in 1910 and a bridge over a dam on the Rhine in 1912. In 1913 he had constructed a hollow-box cantilever-beam bridge at Ibach.

Each of these works has interest in its own right but none is central to the present study. (They all appear on the map and accompanying table in Appendix B below.) One other bridge, however, completed in 1912, was essential to Maillart's later works.

In 1911, Maillart won a competition for a concrete bridge that was to lead him, after World War I, to a new form, the deck-stiffened arch. From four submitted proposals, Maillart's won the competition for a new bridge in Aarburg, over the Aare River. This work, after Tavanasa, was Maillart's only prewar arch bridge that revealed its concrete arch structural form. The three main structural members—deck, columns, and arch—were clearly visible; but unlike the Tavanasa bridge, these members were designed to perform their structural functions separately, independent of each other.

The concrete arch of this bridge (Fig. 6-3) supported very thin concrete columns, which in turn supported both the 5-meter-wide longitudinally ribbed deck and the two solid 1.25-meter-high concrete parapets.[1] The arch was the longest span (67.83 meters) built by Maillart up to that time and had the highest ratio of span to rise (9.75) of any of the cast concrete bridges ever built by his firm. The bridge has a visually striking location at a sharp bend in the Aare River, just below a high bluff commanded by an eleventh-century castle.[2] Given this fine setting, the canton engineer insisted on a handsome structure, ruling in particular that the bridge must be an arch and must have a single span, a pillar in midstream being aesthetically displeasing.[3]

Fig. 6-1. Rhine River bridge at Laufenburg by Maillart, 1911

Fig. 6-2. Rhine River bridge at Rheinfelden by Maillart, 1912

Fig. 6-3. Aare River bridge at Aarburg by Maillart, 1912

The Aarburg bridge helps reveal Maillart's thinking just before World War I. As at Billwil, where the span was about the same length, Maillart took a fairly conventional approach to the design. At Billwil, faced with the need to double the span length of his previous bridge at Zuoz, he had used two spans, each similar to the one at Zuoz. At Aarburg, the program required one span, so he took the conservative line of designing a relatively heavy hingeless arch big enough to carry both the entire dead weight and the complete live load. Thus the Aarburg bridge marked a break away from his earlier practice, from Zuoz to Tavanasa, in which he had designed the deck, walls, and arch to carry the loads as one unit.

On the other hand, Maillart designed extraordinarily thin columns (20 cm × 25 cm) to carry the deck loads to the arch. Visually, these elegant columns contrasted strongly with the comparatively thick solid arch (from 80 cm to 100 cm) and the deep parapet (125 cm). The overall effect was one of two strong members—one straight and one curved—joined by very delicate vertical lines.

These verticals were mere struts to carry vertical loads and were not intended to stiffen either the top or bottom members against bending. Perhaps nowhere else in Maillart's works is there a clearer statement of the structural fact that elements made thinner will carry less bending moment and hence need be only as thick as constructionally practical.[4] Unfortunately, these thin elements deteriorated over time because of water leaking from the faulty drainage system on the deck, which gradually through freezing and thawing caused concrete over the reinforcing steel to crack off, exposing the steel to rust. Thus, by 1968 the columns needed to be rebuilt.[5] We shall return to this and other problems with the Aarburg bridge in the next chapter to see their influence on Maillart's later works.

Origins of the Flat Slab

Despite these bridge projects, the years from 1909 to 1919 were devoted mainly to applications of reinforced concrete to a variety of other undertakings, especially buildings. It was during this period that Maillart developed a new system for building floor slabs on columns without any beams, the flat-plate or mushroom slab. The remainder of this chapter is devoted to this aspect of his work, for although not central to the main focus of our study, it is revealing of his approach to engineering.

In 1912, Maillart published an article on floor construction,[6] his first paper presenting formulas and discussing structural analysis quantitatively. For the historian, it also illuminates his ideas about the design of buildings as distinct from bridges. This paper was stimulated by a public discussion at the Zurich SIA meeting of November 1911. Professor F. Schüle had argued with Maillart by asserting that slabs with no beams were inferior to thinner slabs supported by a grid of beams. Clearly this conversation provoked Maillart, and he explored the question of slab loadings in detail in order to show that Schüle was wrong. But Maillart had another motive, and that was to compare beam-supported slabs with solid ones such as the solid flat-plate slab for which he had just received a patent. Schüle had argued that the extra material was "a useless extra loading," which Maillart disputed numerically.[7] The controversies have long since been resolved, but the two issues that emerged from all this argument are just as valid now as then: efficiency of materials, and safety.

To understand Maillart's work during this period and why he argued in favor of heavier floors, it is necessary to sharpen the basic distinction between a bridge and a building. A bridge structure is essentially a line drawn across some obstacle, and as a result, the problem of bridge form is one of shaping a line: curving down for an arch, up for a cable, and then stiffening it directly by thickness or indirectly by connection to another stiffened line. A building structure, on the other hand, is essentially a set of planes connected together to form boxes that provide shelter for people and materials. In one class of buildings, warehouses, the interior structure is essentially all there is to the building except for the enclosing walls and a few partitions. This quality explains why such buildings appealed to Maillart.

If a bridge is a stiffened line, then a warehouse floor is a stiffened plane. Geometrically, a line is a point moving in one direction, and it is easy to visualize the forces in the bridge moving from, say, a truck at midspan, through the arch, and down to the abutments. In a floor, the problem is more complicated. The nature of multistory buildings requires, in general, flat floors. The roofs may be curved or sloped, but the floors obviously must be flat on top and as flat as possible underneath for convenience of ceiling use.

The fundamental difference in structure between bridge spans and building floors is that for the overall system, the bridge carries its load in a plane parallel to the force of gravity, and the floor carries it in a plane perpendicular to that force. This difference means that bridge structure depends primarily upon geometric form and secondarily upon material thickness, while building structure depends primarily upon material thickness and only secondarily upon geometric form.[8] In more scientific terms, the bridge form seeks to carry its loads in the plane of the materials (the vertical plane), while the building form must carry its loads out of the plane of the floor (the horizontal plane). The structural engineer calls these in-plane loadings *axial* (such as the compression forces directed along the curved axis of an arch) and out-of-plane loadings *flexural* (such as the bending of a floor beam under heavy weight at midspan). Stand a ruler on its end on the table and push down, and you can see how stiff it is under axial loading; but place it horizontally between two books and push down at its center and you can see how flexible it is under flexural loading. The axial member of a structure can be very thin provided it is loaded parallel to gravity, but the flexural members must be thick to prevent breaking.

The form problem in bridging is to make the vertical plane as efficient as possible, which Maillart began to do at Zuoz and improved at Tavanasa. The form problem in flooring is to use the horizontal plane as efficiently as possible, and this is the problem to which Maillart turned after 1905. Bridges, presenting mainly geometric problems, opened up great visual possibilities for the designer, but could not by themselves sustain him in business because the demand for bridges was low. Floors, being primarily a matter of material, opened up great commercial possibilities, but could not provide the designer with much opportunity for visual creativity.

In bridges, the structural action is exposed by geometry, while in floors, it is hidden within materials. Both these results are logical, even though they lead to such drastic differences in visual forms. Sigfried Giedion, in a pioneering study of Maillart, seemed to believe that Maillart's flat slabs never achieved the elegance of his bridges because, in the buildings, he worked with architects of "inferior vision." Giedion neglected the fact that flat slabs are far more difficult to make visually interesting.[9]

Although in his later articles on bridges Maillart stressed lightness and the need to reduce materials, he felt otherwise about floors. Here, materials were to be *added*; the structure was to be made heavier. His academic critic thought the flat slabs had "unnecessary extra weight," but Maillart thought otherwise.[10] His first argument was that in buildings, solid heavier floors were preferable to thin floors supported by ribs underneath, because of better fire protection, better acoustical properties, more resistance to vibration, and less deflection under static loads.

His second idea had to do with load factors. In the early twentieth century, the safety factor for a floor in Switzerland was measured at three times the total of dead load plus live load. Figure 6-4a shows a solid slab with a dead weight of 400 kg/m² and a live load of 200 kg/m², and Figure 6-4b shows a ribbed slab with a dead weight of 200 kg/m² and the same live load. Three times the total design load of the ribbed slab is 1,200 kg/m² and of the solid slab is 1,800 kg/m². Using the total load for the purpose of computing the safety factor satisfied the design profession and general public at the time, but as Maillart pointed out, this was not sensible because it made the heavier slab in effect much safer than the lighter one. The total-load measure of safety unfairly penalized a heavier structure by requiring it to carry not only three times the design live load but also three times its own dead load. Maillart argued that because the dead loads, although different, were constant, the real measure of safety was instead the "safety against live overload." If the dead load were subtracted from the total load of the lighter slab, the remaining live-load capacity represented an overload capacity five times higher than the normal live load of 200 kg/m². Subtracting the dead load from the heavier slab, however, left an overload capacity of 1,400 kg/m², or seven times the expected live load of 200 kg/m².

Maillart thus demonstrated that a greater overload could be supported by a heavier slab with a safety factor of three, or that with a factor of only two-and-a-third,

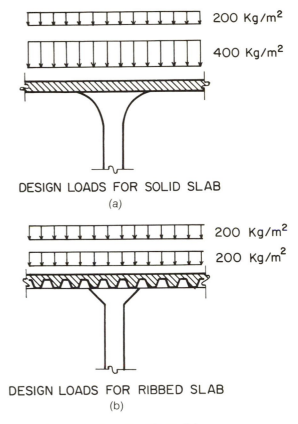

DESIGN LOADS FOR SOLID SLAB
(a)

DESIGN LOADS FOR RIBBED SLAB
(b)

Figs. 6-4a and 4b. Design loads on slabs

his solid slab could support as much overload as a ribbed slab with a factor of three. A Maillart slab with a factor of three was thus worth more money; or, with a lower overall safety factor, his slab could do the same job. Obviously, Maillart had commercial reasons for pointing this out, but in the process he introduced, possibly for the first time, the idea that the safety factor should be split into parts, today called load factors, and applied on the basis of their likelihood (since dead load cannot increase, it therefore should not be multiplied by the same safety factor as should the live load, whose magnitude is always uncertain). Maillart actually proposed load factors that are similar to those put into the American code half a century later.[11]

Behind these justifications for the solid slab lay Maillart's belief that the simpler form work would make those slabs less expensive and hence more competitive. Moreover, because buildings were much more numerous and usually of much greater overall cost than bridges, it was important for a builder to protect his ideas by patents. Hennebique, Wayss, and the American Ernest L. Ransome had partly built their reputations on patents, which were of two types: those for individual components, such as Ransome's twisted bars, tooled joints, and coil-formed lapped reinforcing bars; and those for overall structural systems, such as the "Ransome System of Unit Construction," a method for precasting and connecting together entire buildings, covered by Ransome's 1902 and 1909 patents.[12] The idea of a system was extremely important to builders at the turn of the century. In Christophe's 1902 book, for example, the various major ideas for construction are all described in terms of patented overall systems named for their developers, such as the systems of Hennebique, Monier, Matrai, Bonna, Coignet, and Ransome.[13]

The inventive Swiss had no patent law until 1888 and then only an ambiguous one, until 1907.[14] Their success lay, first, in their ability to patent their inventions elsewhere and thus secure markets in the major countries of Europe; and second, in their skill at imitating foreign inventions without having to pay royalties. Local industries were thus able to compete internationally, especially in dye stuffs, food products (chocolate, condensed milk), and machinery.[15] While there is no evidence that Maillart ever took advantage of Swiss law to use foreign patents, neither apparently did he seek foreign protection for his ideas at this time.

We have already seen that shortly after establishing his own firm, Maillart was granted the ill-fated patent for the arched beam system he had developed for the Zuoz bridge. In his efforts to promote new uses for reinforced concrete, Maillart invested some of his time in a number of inventions between 1904 and 1906. He patented precast reinforced concrete curbstones, notched tramway tracks set in reinforced concrete blocks, and a new way to anchor wooden masts in the ground.[16] While he adapted the tramway system for a later railroad bridge at Liesberg, Maillart apparently was unable to exploit any of these ideas commercially, though they may have had experimental value.

Probably encouraged by the new law of 1907, Maillart subsequently took out several new patents requiring more construction experience than the previous ones and in one case considerable capital. Each was a structural system rather than a component; each had wide application; and each had wide commercial utility, although none possessed as much visual potential as did the three-hinged arch bridges that came from his very first patent. The first was for his flat-slab floor system, the second for arched bridges made of concrete blocks, and the third for a floor system with hollow precast parts. This last one played little role in his future business, but the bridge patent was crucial to his winning the building contracts for the Laufenburg and Rheinfelden bridges and the later Lorraine Bridge design commission in Bern.

The first patent in this second group was for the flat-slab floor and was applied for in January 1909. It was advertised as:

An Important Advance in Reinforced Concrete
Construction
called
The Beamless Deck
System Maillart & Cie, Swiss Patent 46928

From 1910-1912, Maillart completed eleven large projects, almost half of all his work during the period, using this system.[17]

In 1908 he had begun to test models of a floor slab of reinforced concrete supported underneath by a regular grid of isolated columns with no connecting beams (Figs. 6-5 and 6-6). In his first model the slab was connected to four columns by hinged supports. Under testing, the slab bent excessively and failed (see right-hand side of foreground in Fig. 6-5). In his second model, he connected the columns directly to the slab, and then expanded the slab to consist not of one panel supported by four columns, but of nine panels supported by sixteen columns, with the panels all part of one slab. The tests for this slab system were successful enough for him to apply for a patent in early 1909; but, interestingly, the calculations were incomplete.[18] Design again came first and analysis second. Complete analytical jus-

Fig. 6-5. Load test by Maillart on beamless slabs, Zurich, 1908

Fig. 6-6. Nine-panel slab tested by Maillart in Zurich, 1908, with Berthe, Max, and Maria Maillart

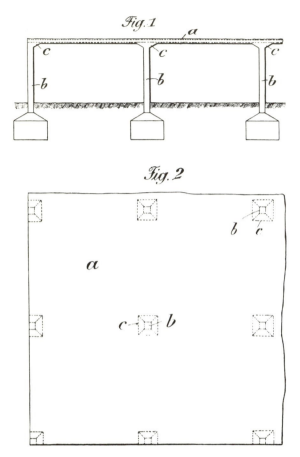

Fig. 6-7. Figures 1 and 2 from Maillart's patent no. 46928, January 1909

tification was certainly unnecessary for the patent description itself, which was remarkably brief; and the accompanying general sketches had no details or relative dimensions (Fig. 6-7). The main patent claims were as follows.

PATENT CLAIM

A space-covering deck structure is to be characterized as a plate reinforced crosswise and resting directly on columns without using any special projecting load-carrying elements running from column to column.

Secondary Claim

A space-covering deck structure following this patent is to be characterized as having the columns, on which the plate rests, thickened at their tops in a corbel-like manner.[19]

There was no mention of spans, thickness, or reinforcement patterns. It was really a concept and not a detailed system. Maillart reported many years later on the results of his early tests (both those of 1908 and later ones of 1913-1914) which are worth describing briefly for the light they shed on his approach to structural analysis.[20] In spite of the simplicity of the patent statement, the structural idea was based in sound research methods.

In the actual tests, Maillart used reinforced concrete on a large enough scale so that the structure could serve as a real building and not merely as a model. And because such a large test structure was costly, he sought as much test information as possible from the single structure. Each of the four exterior sides was designed to provide a different type of support, each typical of the most common wall conditions met in practice.[21] The literature by now is filled with mathematical analyses of plate structures. Some were available even when Maillart began testing in 1908 and a large quantity had been published by the time he wrote up his test results in 1926; but he left those precedents completely aside and concentrated on using his results to develop formulas, which he made up into tables. Maillart's goal was to obtain simple rules for design through these carefully measured practical tests. The time saved in this way could be considerable. As a later associate noted, a mushroom slab floor could be calculated in a half-day or a day and then handed over to Maillart's experienced draftsman.[22]

It is interesting to contrast Maillart's tests of 1908 with those of Arthur R. Lord, who published the results of the first American flat-slab tests in 1910.[23] In his careful measurements on the reinforcing steel, Lord in fact introduced considerable confusion into the understanding of flat-slab structures. Unlike Maillart, who derived the correct amount and location of reinforcement by measuring the vertical bending of the overall slab, Lord measured the strain on an individual reinforcing bar directly and obtained faulty overall results because he failed to include the tension taken by the concrete. By studying the behavior of the complete system, Maillart obtained data that permitted the direct development of simple and correct estimates of local behavior. Lord, on the other hand, by looking only at local behavior, failed to develop his data into accurate estimates for the overall system. When Lord summed up his test numbers, he got a figure that was less than one-third of the correct total, though his results, published in the prestigious *Concrete Engineers' Handbook*, were not corrected until the 1920's.[24]

Not only was the correct amount of reinforcement misunderstood, but the location of such reinforcement was incorrectly perceived. This time the *Handbook* followed the original but unclear ideas of America's eccentric promoter of flat slabs, C.A.P. Turner. Turner neither originated the system nor held any valid patent on it, but he did build flat slabs at least as early as 1906. His book of 1909 established his reputation and introduced the idea that all the reinforcement should lead directly to the columns instead of just going between them.[25] Figure 6-8 shows his pattern of reinforcement which, however elegant, is difficult to build; nevertheless, the *Handbook* supported this pattern and maintained that "it is safe to say that a properly designed flat slab with continuous bands of steel passing over the column heads cannot show a sudden collapse either under test load or in actual service."[26] In addition to the association of proper design with bands of steel passing over the columns, the author of this section of the *Handbook*, Walter S. Edge, gave a "theoretical discussion" of slab behavior, which concluded with a reinforcement pattern of considerable complexity (Fig. 6-9).

Maillart recognized that the problem of reinforcement location was much simpler; the difference between early American practice and Maillart's resided once again in the confusion between analysis and design. To understand this requires further analytic discussion of the differences between bridges and buildings.

A beam bridge loaded at midspan by a big truck must bend to carry half the weight to each abutment. There is no ambiguity in the direction of the load. Maillart's three-hinged arches, like the beam, carry loads in almost this same simple way. But when a large weight is placed at the center of a nine-panel flat slab, it is not nearly so clear in which direction the load travels to get to the column supports, nor is it clear just how much load each column carries.

Fig. 6-8. Four-way reinforcement as used in the United States

Fig. 6-9. Flat-slab reinforcement following the Smulski system

Fig. 6-10a. Flat-slab tests carried out by Maillart at Oerlikon in 1913-1914: overall system showing movable loading

The pioneers of flat slabs for buildings had to develop some method of determining how the load travels in order to provide reinforcing steel bars in the proper amounts and directions. Mathematical methods were both complex and inadequate, and Maillart and others therefore planned to use test results. Tests on reinforced concrete structures measure movements, and the difference between the experiments by Maillart and those by Lord lay in the types of movements each considered primary. Maillart concentrated on vertical deflections, Lord on horizontal strains. Vertical deflections measure the overall behavior of a structure, whereas horizontal strains give local behavior. Figure 6-10 shows Maillart's tests of 1913-1914. Here he used a concentrated load (Fig. 6-10a) at each location, measuring deflections throughout the slab (Fig. 6-10b) by simple wedge devices (Fig. 6-10c).

By seeking first to understand overall behavior, Maillart risked incurring local defects but avoided the risk of an unsafe structure. Lord, on the other hand, by looking first at local behavior, seriously underestimated the slab forces and hence produced design recommendations that were dangerous.[27] Maillart realized early in his career

Fig. 6-10b. Flat-slab tests carried out by Maillart at Oerlikon in 1913-1914: underside of slab showing measuring points

Fig. 6-10c. Flat-slab tests carried out by Maillart at Oerlikon in 1913-1914: detail of measuring device

that reinforced concrete was a special material of which the local properties were so uncertain that the only sensible approach to an understanding lay in looking primarily at the whole structure. Local cracks in concrete structures are inevitable and the design problem is to control them, not to assume they do not exist. Lord did not recognize the relative unimportance of local measurements and hence, by assuming that the concrete behaved the same everywhere, he seriously misunderstood his results. His primary interest in local effects made it difficult for him to see the overall behavior. This problem fits well the cliché of not seeing the forest for the trees.

Maillart's decision to observe the entire structure came after his first (1908) test failed because of a faulty concept. He had learned from it just as he had learned something when his first hollow-box, three-hinged arch bridge at Zuoz exhibited cracks because of a local defect. His ideas grew as his experience with structures under load broadened. It was his reflective interpretation of defects that led to more mature ideas and ultimately to his most stunning works, in which visual elegance is fully matched by the beauty of the purely technical underlying principles.

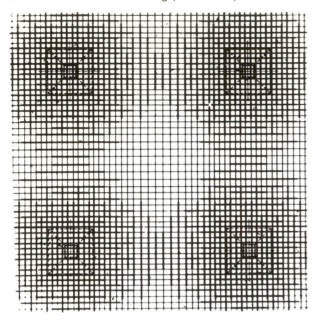

Fig. 6-11. Two-way reinforcement as used by Maillart

Fig. 6-12. Giesshübel warehouse in Zurich by Maillart, 1910

(a)

(b)

Figs. 6-13a and 13b. Fabrik Prowodnik in Riga by Maillart, 1914: (a) exterior view during construction (b) roof support

For these early flat slabs, Maillart recorded the vertical deflections during the tests and turned them into a simple set of formulas. On the basis of these he designed a simple reinforcing pattern (Fig. 6-11) that has since become the accepted design for flat slabs in the United States and elsewhere. He had already seen this solution in 1910 when he built his first major flat-slab building, the Giesshübel warehouse in Zurich, still in excellent condition sixty-eight years later (Fig. 6-12). Maillart's economical and attractive flat-slab structures were welcomed not only in Switzerland but in the developing economies of the European periphery. A flat-slab warehouse he built in St. Petersburg in 1912 led directly to the beginning of a large Russian branch of his firm, while the cable factory he built in Villanueva in 1914 drew enough new business to justify a Barcelona office that had completed a series of large buildings by 1920.

The Maillarts planned to spend the summer of 1914 in Riga to take advantage of the noted beach resort nearby while Maillart worked on the large factory in reinforced concrete that he was building there (Fig. 6-13). When war broke out, the family went to St. Petersburg and eventually to Kharkov, where Maillart had another huge project. They were then effectively trapped in Russia, where Maria died of a gallbladder infection in August of 1916. Finally the Bolshevik revolution forced Maillart to leave in late 1918. When he returned to Geneva at the beginning of 1919, he began a new life, imposed on him by the tragedy of losing his wife and the professional dislocation of losing his business. He did not remarry. Never again would he have his own house, and never again would he build his own designs. Yet his design ideas showed a remarkable continuity as he returned to work and began to reestablish himself. His very quick professional recovery was based on his earlier building experience; and it is to this recovery that we next turn.

Fig. 7-2. Magazzini Generali in Chiasso by Maillart, 1924

cost. What they do signify, however, is a more efficient use of materials by Maillart in the capitals.[3] Since that efficiency had little to do either with performance or cost, he probably chose this design for its appearance. As he later wrote when comparing the American system to his own, "the more rational and more beautiful European method of building" would eventually be accepted everywhere.[4]

The roof of the shed adjacent to the warehouse revealed similar concerns. Whereas the beamless slab was a sensible engineering solution for relatively heavily loaded floors with columns spaced in regular, nearly square grids, the gabled truss (Fig. 7-2) for the Chiasso shed was a sensible engineering solution for relatively lightly loaded roofs, with supports spaced in a regular but rectangular grid where the span length greatly exceeds the truss spacing. The long-span roof has the same basic characteristics as a long-span bridge, where the major structure, being in one vertical plane, becomes more a geometrical design and hence could provide Maillart with a chance to think out a new visual problem. Like his original bridge at Zuoz, the roof displayed not a straining for new effects, but a contemplation of old forms, used in new ways. Maillart seems to have had two sources in mind, one rational, the other visual.

7 From Builder to Designer (1919-1927)

Returning to Switzerland in early 1919, Maillart had to restart his professional life.[1] His contracting business having virtually ceased by the end of the war due to his absence from Zurich and his financial losses in Russia, he decided to pursue the less strenuous career of a consulting engineer rather than return to building. It was a hard time to change, but in this second phase of his career he designed all the major works on which his fame now rests. Three projects characterized this change.

First, there was a new warehouse and shed for the Magazzini Generali at Chiasso in 1924, combining flat-slab construction of the major structure with a unique design idea for the roof of the adjacent one-story shed. The second project was a series of small bridges around the Wäggital Lake, designed between 1923 and 1924. The third was the Lorraine Bridge in Bern, which allowed him to establish his bridge office in the Swiss capital. In each of these three projects, Maillart reworked ideas from his early building experience; each can in fact be traced to one of his major prewar works.

A New Form in Buildings

Maillart naturally turned quickly to designing large buildings, using his invention of 1909. He designed the Chiasso warehouse just as his patent expired. It was the first Swiss flat-slab structure to be load tested publicly.[2] Between 1925 and 1930 Professor Mirko Roš tested eight flat-slab structures built in France and Switzerland, of which three were by Maillart. Roš's report lets us compare Maillart's work in the late 1920's with that of his competitors.

Figure 7-1 shows the column capitals for six of the eight structures described by Roš; the other two were Maillart designs similar to the 1925 one shown here. Whereas Maillart's design provided a smooth transition between column and slab, all the others achieved the structurally necessary column widening by discontinuous breaks in profile. Only the Locher design of 1930 used a smooth capital, but its transition from a square plan capital to a hexagonal plan column led to discontinuities of profile at the column.

These differences do not noticeably influence the overall structural behavior of the slab-column structure and probably they did not appreciably affect the construction

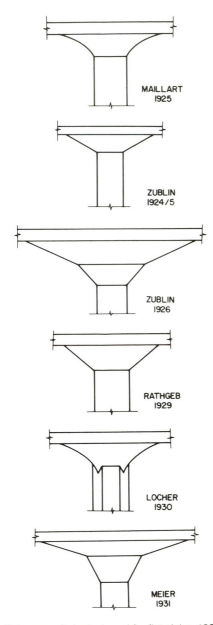

Fig. 7-1. Column capitals designed for flat slabs, 1924-1931

Fig. 7-3. Karstelenbach bridge at Amsteg on the Gotthardline by Gerlich and Brack, engineers, 1881

Fig. 7-4. Karstelenbach bridge as strengthened in 1908

Fig. 7-5. Mittlere Maienreuss bridge near Wassen as strengthened in 1908

As with the bridge designs a quarter of a century earlier, the first aspect to explore is the visual one. Just as the Stauffacher form suggested the Zuoz rationale, so another late nineteenth-century form, this time associated with railroad bridges, seems to have suggested to Maillart the form for the Chiasso shed. These older visual forms were to be found on the Gotthard line completed in 1882.[5] The most impressive section of this line occurs in the upper Reuss valley between Amsteg and Göschenen in the old inner Canton of Uri. Here the railroad must rise 582 meters in only 16 kilometers, passing through three loop tunnels and over a series of iron-truss, masonry-supported bridges. Contemporary Baedeker guides, which as a rule did not note engineering projects in regions of such breathtaking scenery, devoted most of their landscape description to these works, calling the bridges "handsome" and "imposing."[6]

In 1908, these bridges were strengthened for heavier traffic, enhancing their already striking appearance. A metal inverted arch constructed below the latticed truss changed the constant-depth truss (Fig. 7-3) to one of variable depth (Figs. 7-4 and 7-5) so that it became almost twice as strong with the addition of very much less than half the original material.[7]

It was over this line that Maillart repeatedly traveled in the years 1908 to 1914 and again between 1920 and 1923 prior to his design work on the Chiasso project; and these bridges, unique in appearance, were the most visually impressive feats of modern engineering on the direct route between Zurich and Chiasso. Although there are many bridges designed with trusses of this same general shape, these Gotthard line works are among the very few that have the simple lines of a hanging polygon connected by widely spaced thin vertical elements going to the top member. If the heavy trussed deck is replaced by a wide roof and if the entire system is kinked upward at the middle to provide for snow slopes, then the 1908 bridge forms on the railway to Chiasso become the rail shed form in the Magazzini Generali of 1924.

There is also a rational basis for the form of both the bridges and the shed roof.[8] That form can be characterized by a diagram of the critical internal force distribution across a uniformly loaded beam (Fig. 7-6). This diagram shows that the loads will bend the beam downward over the central length (+ region), causing tension in its bottom part and compression in its top. Where the diagram crosses from + to −, there will be no bending, and in the outside lengths b the tension is on the top and the compression on the bottom. The vertical distance on the diagram in Figure 7-6 represents the intensity of the force, i.e., a maximum positive at midspan, zero at the crossings, and a maximum negative over the supports. If the designer makes a geometric truss form shaped like the diagram, then the major internal forces will be nearly constant throughout the truss length, a condition permitting the longitudinal truss members to be the same size. Figure 7-7a shows such a truss in which the members labeled T are in tension and those marked C are in compression. Finally, when the truss is gabled to prevent heavier snow loadings, the result (Fig. 7-7b) is the Chiasso shed truss form.

The form presented two further problems because of Maillart's desire to use a minimum of materials. First, since the top compression strut, C_1, was very slender, it would be overstressed; so Maillart designed it to be monolithic with the concrete roof slab, thus giving that slab—already needed to carry loads transversely between trusses—a second use, that of carrying longitudinal compression. Second, the large compression forces in the slant elements, C_2, between the zero point and the support, required additional material. Maillart again provided this by designing the transverse frame between supports to be monolithic with the element C_2, again giving a necessary element—the transverse frame—a second structural function. (See the splayed-out forms at the top of the column in Fig. 7-8.)[9]

The detailed structural design followed an idea central to reinforced concrete construction, even though the suggestion of the form may have come from a metal bridge. That idea is simply that concrete material is used both to resist compression

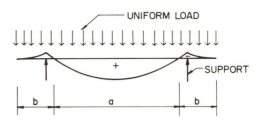

Fig. 7-6. Bending moment diagram for a beam with overhanging ends

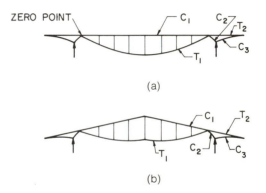

Figs. 7-7a and 7b. (a) Constant force truss (b) Gabled constant force truss

and to protect the tension-carrying steel from corrosion. Thus, the bottom polygonal members need not be nearly so large as the top gable because those bottom members are essentially steel (similar to the bottom members of the Gotthard line bridges) and hence the concrete area can be very small.[10]

If the visual inspiration came from specific metal bridges and the rational basis from the theory of structures, there was one final determinate of form that must have underlain both, namely, Maillart's idea about what was appropriate for concrete structure. Again, understanding this idea requires a further excursion into the origin of structural form. Just at the time of Maillart's first bridge works, a professor at the University of Louvain, A. Vierendeel, had published his early ideas on trusses without diagonals[11]—essentially the principle used in the Chiasso shed. Maillart was undoubtedly aware of this new form for concrete trusses, and the contrast between the forms that came out of Vierendeel's idea[12] and those of Maillart is very important.

When Vierendeel worked on his design, he always had before him the image of a single element, the truss made structurally incomplete by removing the diagonals, so that the truss forms had to be stiffened by using heavy columns and large rigid connections (Fig. 7-9). Maillart had the image of two elements, the gabled roof in compression and the inverted arch in tension, made complete by adding the vertical connections of the struts. In the one, the visual dominance of the missing diagonals within the form made the verticals alone seem too slight, too uncertain, indeed too unsafe; whereas in the other the visual dominance of a single overall geometric form made the verticals seem to act as necessary additions.[13] In the one they symbolized uncertainty and incompleteness; in the other, clarity and integrity. If there is any doubt about this difference, one need only compare typical Vierendeel bridges in concrete with deck-stiffened bridges, to which we shall return later in this chapter.

The Chiasso shed provides our first major example of Maillart the designer putting into real form his mature ideas, which had developed slowly over thirty years of engineering work. It displayed his conception of the whole preceding the parts, of simple as opposed to complex shapes or connections, and of evolution rather than novelty for its own sake. In the shed, the form had an overall gable-parabolic shape; it had everywhere simple, thin elements; and it was inspired by existing bridges of the region.

Equally important for the development of Maillart's ideas was a problem that arose from a fault in the design of the Aarburg bridge. Apparently as early as 1915, large cracks appeared in the deck just above the columns.

Fig. 7-8. Underside of the shed roof for the Magazzini Generali at Chiasso by Maillart, 1924

Fig. 7-9. Landquart River bridge near Dalvazza by N. Hartmann & Cie, 1925

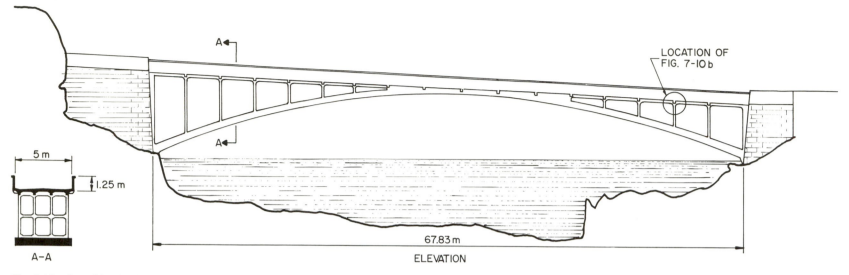

Fig. 7-10a. Aare River bridge at Aarburg by Maillart, 1912

Fig. 7-10b. Longitudinal deck beam near quarterspan of the Aarburg bridge showing vertical cracking

A New Form in Bridges

Like other prewar bridge engineers, Maillart had not fully understood the interaction between the arch and the superstructure. Wide cracks had occurred in the Aarburg bridge of 1911 over the supports owing to vertical deflections in the longitudinal (deck) members,[14] for although the Aarburg bridge had the appearance of Maillart's later deck-stiffened arch bridges where the straight deck has a deep parapet wall, it lacked the internal reinforcement of the later bridges (Figs. 7-10a and 10b).[15] The difficulties with this bridge, as well as the clear visual image of a stiff deck, probably led Maillart to think about the interaction of deck and arch in a new way after the war.

It was not until 1923 that Maillart picked up again the line of thinking begun with Zuoz, and converted it into a new deck-stiffened form for the Flienglibach bridge (Fig. 7-11a) high up in the obscure Wäggital in Canton Schwyz. The idea is relatively simple, and consists of designing a stiff longitudinal parapet that serves as a straight deck-girder and is connected through slender transverse cross walls to a thin arch below the deck (Fig. 7-11b). The stiff parapet prevents the arch from bending under heavy traffic loads, and thus permits the use of an arch as thin as can be accurately built.

Figure 7-12 shows the first two figures from Maillart's 1902 patent, which had already set forth his major bridge ideas. The first figure, an elevation, shows the solid wall girder similar to Stauffacher (in appearance only), Zuoz, and Billwil; while the second figure, a section, shows the deck and the arch connected by transverse walls. Of course this second figure did not mean in 1902 that the transverse walls were the essential connectors of the deck and arch; Maillart clearly stated in the patent that the connection was by the longitudinal walls, which in a section view appear only as the background part labeled *a*.[16] However, that section view does clearly suggest visually that the transverse walls could provide the main connection between deck and slab. Maillart seemed even to imply that possibility in his patent statement: "The way in which the connecting steel bars are arranged is unimportant; what is essential is merely that the three major structural elements be well connected together. They can thus be disposed in other ways than that shown in the figure, or even built completely as a framework."[17]

One possible interpretation of that last phrase "built completely as a framework" would be to design the deck, the transverse walls, and the arch as a framework much like the Vierendeel truss. One further truss-like implication arises from the two openings shown in that second 1902 figure and noted in the patent statement: "In

(a)

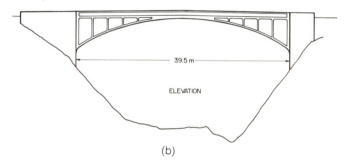

(b)

Figs. 7-11a and 11b. (a) Flienglibach bridge by Maillart, 1923, under construction (b) Flienglibach bridge

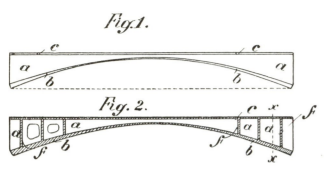

Fig. 7-12. Figures 1 and 2 from Maillart's patent no. 25712, February 1902

case openings are needed in the walls, they can be provided (Fig. 2, left), as can stiffening walls in the transverse direction."[18] Therefore, Maillart had already considered the possibility of reducing the walls and, as the left side of the figure suggests, leaving mainly the transverse walls as connectors. The Tavanasa idea of eliminating the end section of the wall is also implied by these openings, but visually the figure is closer to Flienglibach than to Tavanasa. In sum, the *Bogenträger* idea contained the seed for the deck-stiffened concept; first, by its technical rationale of connecting the deck and the arch together to save materials; and second, by its visual suggestion of that connection by transverse rather than by longitudinal walls.

However, the 1902 figure did not give any indication of the necessary relationship between the stiffness of the deck and the arch, because at that time Maillart saw the longitudinal wall as the stiffener for the thin arch. The strong visual suggestion provided by the deep parapet and slender columns in his own Aarburg bridge of 1911 and the observed cracking in its underreinforced deck could well have brought him directly to his new form when given an occasion in 1923 to consider arch bridges from the perspectives of efficient use of materials and economy in construction costs. The hydroelectric project in the Wäggital provided him with this opportunity.

In October 1921, the city of Zurich along with the Power Company of Northeast Switzerland created the Wäggital Power Company, which was to build a dam across the Wäggi River and a hydroelectric power station. The Wäggital project was one of the largest works under construction in Switzerland just at the time Maillart most needed work.[19] His first major design in the Wäggital appears to have been the aqueduct over the Trebsenbach, completed in November 1923.[20]

His association with the aqueduct builder Simonett and his firm gave Maillart the chance in the summer of 1923 to think again about bridges, as in July Simonett began building the roadway along the right bank of the newly made Wäggital lake.[21] This road crossed a series of small brooks that had to be bridged and posed three types of structural problems. First, and of little importance here, were very small crossings that Maillart made by flat ribbed slabs; second, there were larger brooks in deep ravines to be crossed directly by the road, the first being the Flienglibach; and third, there was one brook (Ziggenbach) in a ravine so wide that the road had to make a U-turn to arrive at a narrower section where the bridge span would not be too great.

For the Flienglibach, Maillart sketched a design which was drawn up on August 2, 1923 (Fig. 7-11b) and bore a strong resemblance to a drawing by Vierendeel.[22] The similarity is more in the initial sketch than in the basic idea. For Vierendeel, this was a truss without diagonals; he saw the vertical members as having to carry not only vertical stresses but also those horizontal stresses ordinarily carried by the diagonals. Thus the analysis became complicated and his designs show heavy connections between verticals and the top and bottom members in order to reduce these horizontal stresses. By contrast, Maillart saw the system as an arch and a deck that deflect to the same extent in the vertical direction because connected together by the columns. In his calculations for the Flienglibach, he said he would design the system in such a way that the arch would carry the load spread evenly over the bridge, that is, the dead weight and a snow load; and for uneven loads, the deck would prevent the arch from bending, so that they too could be carried by the arch as if evenly distributed.[23]

The following year he designed the Schrähbach bridge on the opposite bank of the Wäggital, and in 1925 came the Valtschielbach bridge near Donath, south of Thusis in the Graubünden.[24] The Schrähbach bridge was his sixty-fifth project after his return to Geneva following the war and the Valtschielbach his ninety-fourth; between the beginning of the Wäggital project in 1922 and the completion of Valtschiel, Maillart had succeeded in establishing himself once again and had begun to build a new reputation. Only one further set of events was needed to confirm that reputation and to account for the flowering of his work over the last fifteen years of his life. Those were the events that led to the establishment of his bridge office in Bern.

A New Form of Business

In 1911 the city of Bern had organized a competition for a major new bridge across the Aare River between the old town and the northern suburb of Lorraine. Maillart, along with the Bernese architectural firm of Joss and Klauser, had entered this competition with a concrete arch, but had lost to a masonry design, failing to win any prize.[25] Apparently because of the war, plans were suspended until 1923, when the city again took up the project seriously. At that time, the building commissioner and the city engineer approached Maillart, impressed with his previous design for the bridge, and requested a new study.[26] Maillart completed a new design by 1927 in collaboration with the architects Klauser and Streit (successors to Joss and Klauser). Construction began the next year, and the bridge was dedicated in May 1930.[27]

In retrospect, Maillart regarded the Lorraine Bridge as a commercial success, one of his "beaux morceaux," as he put it, as opposed to his design achievements.[28] The

Fig. 7-13. Lorraine Bridge over the Aare River in Bern by Maillart, showing construction of the central arch ring out of concrete blocks, December 1928

form was traditional; he acknowledged a debt to Bern's Nydegg bridge of 1844. Yet his means were original and economical. He built the central width of the arch first, directly on the scaffold. Once complete, that block arch, built with the blocks in crenulated arrangement, was able to support the bands of blocks placed on either side (see Fig. 7-13). In this way Maillart could make a much lighter scaffold that had only to support the central band (Fig. 7-14) and was hence a less expensive structure. He had previously patented this system, apparently first in Germany for the Rheinfelden bridge and then in Switzerland.[29] The idea of lightening the scaffold derived directly from Steinach and Solis, even though the method in this case consisted of horizontal bands rather than the concentric vertical rings in the earlier works.

Thanks to the Lorraine Bridge contract, Maillart was able to open a Bern office in 1924. He found a lifelong and highly capable associate in Ernst Stettler, who be-

Fig. 7-15. Lorraine Bridge, completed 1930

Fig. 7-14. Lorraine Bridge showing light scaffold with arch in place, March 8, 1929

came his chief bridge engineer and head of this office in 1926; and he acquired a reliable business partner in the bridge's general contractor, the Bern builder Eugen Losinger. Working with Losinger and with the firm of Prader in Zurich, Maillart again could practice in effect as a designer-builder.

But even as Maillart stood before the Bernese people in May of 1930 to speak about the Lorraine Bridge (Fig. 7-15), a far greater work was drawing to an end on the other side of Switzerland, a bridge that differs so radically from the Lorraine that it hardly seemed possible they could have the same designer. That bridge, over the Salgina brook in a high mountain valley of the Graubünden, demands that we explore now the final phase of Maillart's transition from inventive nineteenth-century engineer to the most radically modern structural designer in the first half of the twentieth century.

Fig. 8-1. Valtschielbach bridge near Donath by Maillart, 1925

8 Tavanasa, Salginatobel, and Beyond (1927-1940)

The Lorraine was the last of Maillart's masonry-like bridges, while the Salginatobel was the first of his thoroughly concrete bridges. In the Lorraine there is almost no trace of the twentieth century—as Maillart himself said, the design reflected Bern's Nydegg bridge of 1844—while in the Salgina crossing, he broke definitively with nineteenth-century ideas of design.

Nevertheless, the origins of the Salginatobel bridge are to be found in Maillart's earlier experiences, particularly (as with the origins of the Tavanasa and Flienglibach bridges) in his firsthand observations of previously completed works, including the physical behavior of deflections and cracking on the one hand, and the aesthetics of form on the other. One particular event appears to have been critical to the mature development of the flat, hollow-sectioned, three-hinged arch bridges completed during the last decade of his life.

In September 1927, a landslide swept down and tore out the 1905 Tavanasa bridge over the upper Rhine River.[1] The most daring Swiss bridge of its time was reduced to a pile of debris on the left bank. And yet, the evidence points clearly to the liberating influence this misfortune had on Maillart's work. Between the destruction of Tavanasa and the completion of Salginatobel, his designs took on an additional element of audacious vigor that expressed his full maturity as an artist.

Tavanasa

From a purely scientific point of view, the disaster on the upper Rhine provided a unique opportunity to test the materials in a relatively old concrete bridge. Mirko Roš, after making the tests, concluded that the bridge was well built and had been in good condition after twenty-two years of service in a harsh climate. He went on to declare it "a small but exemplary masterpiece of Swiss bridge building. Its remains are now silent witnesses to the ability and the love of the designer for his work."[2] He had become a close friend of Maillart's and his admiration clearly shows through. But apart from private friendship, Roš's feeling came from professional respect. Building upon the accumulating results of the detailed tests he had been making on Maillart's works, he praised their "technically efficient ideas and artistic character" in a major article.[3]

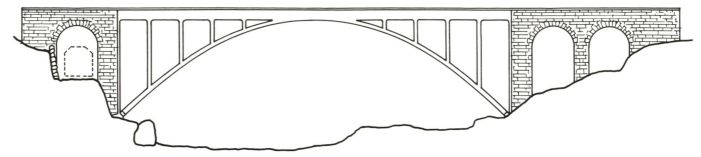

Fig. 8-2. Maillart's design of November 20, 1927 for the new Tavanasa bridge

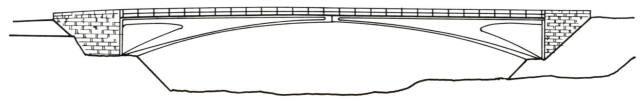

Fig. 8-3. Maillart's design of December 6, 1927

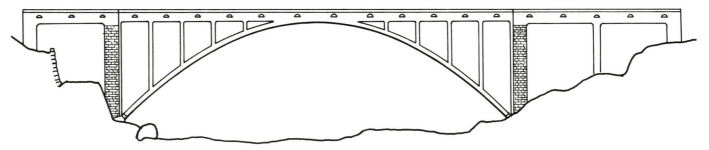

Fig. 8-4. Maillart's design of January 31, 1928

Maillart had begun to work on two designs for a new competition, one to rebuild his earlier bridge, and a second to replace a surviving wooden bridge. Of the second, one drawing remains, dated November 22, 1927, and shows a design almost identical to the 1905 Tavanasa except that the span is only 47 instead of 51 meters. Of the first design, two drawings are left, dated November 20 and 22.[4] These drawings show a relatively thin 50-meter-span concrete arch with a rise of about 13 meters. Eight relatively slender cross walls connect the arch to the deck, which is made of deep parapets and is merged with the arch over the central part of the span (Fig. 8-2). In short, the design is a direct paraphrase of the Valtschielbach structure completed the previous year barely thirty kilometers away (Fig. 8-1).

The town of Tavanasa encouraged Maillart's entry for the replacement bridge. The ground rules of the competition, set by the canton, had recommended a high bridge, with a roadway about 13 meters above high water, over twice the height of Maillart's 1905 bridge; while the town variation was for a lower bridge, similar to the previous one.[5] It appears that Maillart worked on this latter variation, as two drawings dated December 6, 1927 show almost a replica of his earlier Tavanasa bridge (Fig. 8-3). Since these were prepared after the competition deadline, it seems that the town, probably interested in a less expensive structure, and possibly because the people had become attached to his previous design, asked him to make a new bridge along the same lines as the former one.

But the canton reaffirmed the decision for a high bridge in January. Maillart apparently returned to his Valtschiel-type proposal and produced on January 31, 1928 another variation, changing only the approach spans. Lower cost was not the only reason. With a relatively slender reinforced concrete straight-girder framing, Maillart, for the first time in a major project, introduced an approach span that had almost the lightness of his mainspan structure (Fig. 8-4). In his efforts to win the second Tavanasa, he had removed one more vestige of masonry. Still, he left one stone pier separating the arch span from its approaches, and the design had a small but visually upsetting discontinuity, because the longer approach girder was slightly deeper than the deck stiffener over the arch. A final small change in the January 31 drawing was the addition of semicircular openings in the parapet, just as at Valtschiel. As usual, Maillart was approaching a major new design idea at a slow, halting pace, where the power of past images slowed the shift to a new vision; but he was constantly reaching for a new result that would be both more rational and more beautiful, and not merely one or the other.

Fig. 8-5. New Tavanasa bridge by W. Versell, 1928

The competition had brought together all of Maillart's previous collaborators in the Graubünden: Westermann, with whom he had built Zuoz; Hartmann, the builder of Valtschielbach; and Prader, his present contractor.[6] In March, the executive council of the Graubünden accepted the advice of the canton building department and gave the project to the Chur firm of Caprez.[7] The resulting bridge looks very much like Maillart's design of January 31, except that it has only four cross walls between deck and arch, whereas Maillart had eight (Fig. 8-5), and the Caprez bridge was not designed as a deck-stiffened arch.

Still, Maillart's connection with Tavanasa did not cease. With a geologist, he soon received a commission from the canton to evaluate the foundations for the new bridge,[8] and another to represent them in planning with Professor Roš the full-scale load tests in July.[9] Consequently, as the new bridge rose, Maillart followed it closely. The ideas first developed in the Tavanasa competition would soon be applied in a much larger and far more difficult structure, the crossing of the deep Salginatobel.

Background to Salginatobel

In 1914, the little village of Schuders had brought to the cantonal assembly its request for a better route to the town of Schiers; the existing one involved a hike of at least two hours along a narrow road on the edge of a sliding mountain.[10] Nothing was done until after the war, and not until 1928 did construction begin on the road from Ottenacher to the Salginatobel.[11]

In the summer of 1928 as a new road progressed up from Schiers, and as the new Tavanasa bridge neared completion, the canton announced a new competition for a bridge to span the valley of the Salgina brook.[12] The cantonal engineer met with the town leaders of Schiers to discuss the project.[13]

On the deadline of September 15, 1928, Prader & Cie offered to the building department of the Canton of Graubünden

Fig. 8-6a. Partial elevation for the Salginatobel bridge, Variation I

> to build the bridge over the Salginatobel above Schiers and further to do it for the lump sum of: alternative 1 = Fr. 135,000; alternative 2 = Fr. 144,000.
>
> The plans have been made by the engineer, Mr. Maillart, in Geneva and we refer you especially to the enclosed illustrated report.
>
> During the construction, the noted reinforced concrete specialist and designer, Mr. Maillart, will work closely with us, bringing us his rich experience. . . .[14]

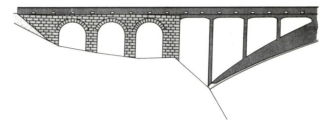

Fig. 8-6b. Partial elevation for the Salginatobel bridge, Variation II

In addition to the report, Prader sent a cost estimate, a perspective view, several sections, and elevations of the two alternatives (see Figs. 8-6 and 8-7). These drawings, dated August 31, 1928, were strikingly similar to the two variations for Tavanasa drawn in November of 1927 and January of 1928 (see Figs. 8-4 and 8-2). Indeed, we can construct the Salginatobel proposals directly from those for the new Tavanasa. All features of the later proposals were present in the earlier competition drawings: first, the open three-hinged arch form (Fig. 8-3); second, the open slender cross walls connecting arch and deck (Figs. 8-4 and 8-2); third, the concrete parapet with semicircular openings (Fig. 8-4); and finally, the alternative approach designs—one with Roman arched masonry (Fig. 8-2) and the other with open rectangular concrete framing (Fig. 8-4).

There is not a single new element either technical or visual, and yet as put together in Geneva on August 31, 1928, this design created perhaps the most stunning concrete bridge of the twentieth century (Fig. 8-8). During that summer between the loss of the Tavanasa contract and the competition for Salgina, Maillart, as he had done before, borrowed elements already used in other works and put them together in a

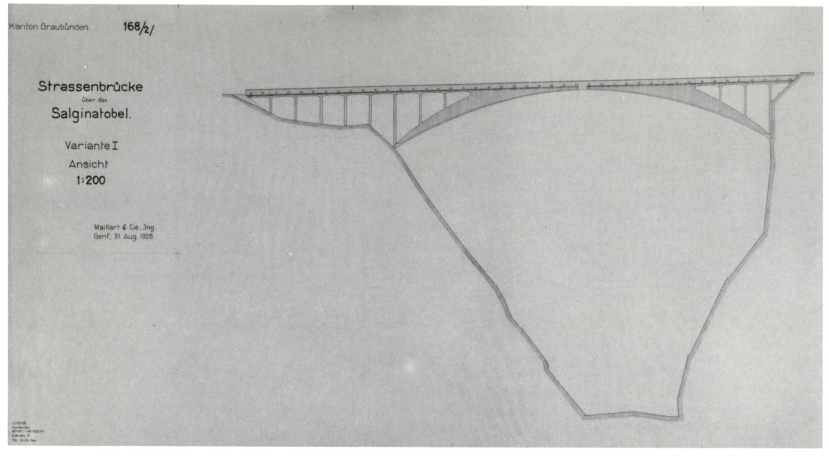

Fig. 8-7. Original drawing for the Salginatobel bridge by Maillart, August 31, 1928, Variation I

new form. He would continue to develop this design until the end of his life so that, as Giedion has said, "Maillart's most important works . . . were nearly all accomplished during the last ten years of his life. As he advanced in age, his bridges became more daring in aspect and more imbued with the vigor of youth."[15] It is difficult to escape the conclusion that without the stimulus of the destroyed Tavanasa, there would have been no Salginatobel, at least not in the form it took during the late summer of 1928.

Fig. 8-8. Salginatobel bridge by Maillart, 1930

Salginatobel

The Salginatobel bridge, built almost exactly as drawn in Prader's alternative 1 came directly out of the Tavanasa studies. These designs formed a curious mixture of Maillart's immediate 1926 past at Valtschielbach and his distant 1905 past at Tavanasa, which in turn came from his first major work, the Zuoz bridge of 1901. All of these bridges are in the eastern Canton of the Graubünden. In a literal sense, the Salginatobel bridge is the direct product of its designer's Graubünden experience. Although Maillart's bridges are found in many parts of Switzerland, if one had to pick just one canton in which to see them, it would unquestionably be the Graubünden. Of Maillart's closest professional friends, Simonett, Carl Jehger, Florian Prader, and Simon Menn were all Bündners. His main office in Geneva was directed by Meisser, also a Bündner. Even though Prader's firm was based in Zurich, to the Graubünden officials it was a Bündner firm.

The Graubünden has perhaps the oldest and longest tradition of direct democracy in the western world. This local democratic experience influences both the way Bündners think and the way they build.[16] They realize, as the rapid rebuilding of the Tavanasa bridge illustrates, that artificial structures are essential to their lives; and yet they consider more than just their use for transport or profit. The practice of the elected town leaders sitting together with the canton engineers and debating the merits of various bridge proposals demonstrates local opinion influencing central authority. If the parties always compromise on the results, they at least agree that public works should come from public decisions as well as specialists' designs. Nearly all of the spectacular bridge works in the canton have an indigenous character—from the masonry arches for the Rhätische Bahn (see Fig. 3-4) to the concrete arches and prestressed spans of the modern Bündner, Christian Menn (Figs. 8-9a and 8-9b).

The district engineer P. Lorenz's report and recommendation still exist in cantonal files and clearly explain the basis on which Lorenz chose the Maillart design for Salginatobel over the other eighteen designs submitted.[17] None of the seven steel projects included any cost accounting for maintenance (painting), so that when Lorenz added 25 percent to the initial cost, the lowest bid in steel came to 172,500 francs or 37,500 francs higher than the lowest bid of the twelve in concrete, namely, Prader's bid of 135,000 francs for alternative 1. The low bid, combined with the high reputation of the builder and the fact that Prader was from the Graubünden, made the choice easy.[18] Three decades of experience with reinforced concrete in the canton

Fig. 8-9a. Nanin and Cascella bridges near Mesocco by Christian Menn, 1967

Fig. 8-9b. Salvanei bridge near Mesocco by Christian Menn, 1969

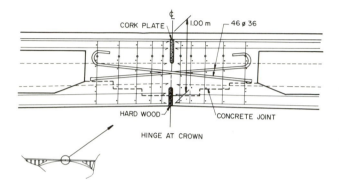

Fig. 8-10. Hinge at crown, Salginatobel bridge

had demonstrated its low maintenance requirements; and the recent tests on the remains of the destroyed Tavanasa bridge had shown its qualities of endurance. The railroads, which traditionally preferred metal structures, were building reinforced concrete bridges and even rebuilding steel work in reinforced concrete, such as the Grandfey viaduct between Bern and Freiburg, also a joint project by Maillart and Prader.[19]

Finally, Lorenz reviewed very briefly two other designs in concrete, one by G. Lüscher and one by Westermann, who again apparently had a very low bid. His objections to these two are less important than his dismissal of the Maillart-Prader alternative 2: "The alternative of the Prader design with a massive approach viaduct in round arches may be offered as a concession to more usual forms; but because of the lack of attractive building stone, it cannot be seriously considered."[20] The primary objection was neither one of cost, the difference being small anyway, nor one of aesthetic rejection of massive form per se; but that proper stone was lacking. Still, there was a secondary or implied objection to the masonry in that it was more a concession to the usual than an outgrowth of the structural requirements. Engineer Lorenz brought no strong modern ideology of form to his judgment but neither was he constrained by tradition. In fact, he went on to define specifically why Maillart's new form was superior to the others, first by grouping it together with a second design and then by explaining its superiority over even that one.[21]

Only the designs of Maillart and Lüscher were appropriate to the sharp vertical drop and relatively weak supporting capacity of the Salginatobel rock walls. A deeper arch would have required very much longer piers or cross walls between arch and deck and a larger amount of concrete, while an arch made heavier at its supports would have necessitated more rock excavation. The Maillart design, with its arch and main cross walls widened to six meters at the support, was more stable, according to Lorenz, than the Lüscher design.

The executive council of the Graubünden awarded the Salgina project to Prader & Cie in December 1928. Maillart's Geneva office completed all the official drawings by June 1929, and Roš reviewed them. In his report, Roš recommended a series of relatively minor modifications, all in the direction of more materials, mostly steel reinforcement. Maillart defended his design but agreed to add the additional steel provided the canton would pay the extra costs. This minor controversy between two very close friends is of interest, not in technical detail, but rather in showing the way Maillart defended his design. Taking only the first point raised by Roš, we can illustrate Maillart's reaction. The issue was the reinforcement through the three hinges of the arch (Figs. 8-10 and 8-11). The idea of the three-hinged arch centers on its ability to adjust to the small but measurable movements in the mountain. Where an arch without hinges will crack as the foundation moves, one with hinges adjusts through the rotations permitted by the hinges (Fig. 2-3). The hinges serve, in effect, as built-in cracks, similar to those in a concrete pavement.

The problem arose at Salgina because these hinges, especially at the foundations, had to be strong enough to carry the entire bridge load into the supports, but flexible enough to permit rotation and thus prevent other parts of the arch from cracking. Maillart's solution consisted of a sharply reduced concrete hinge section through which steel bars were passed. Thus, the small remaining concrete section had to be highly stressed for the hinge to work. This is the so-called concrete hinge, invented in France by Mesnager and Freyssinet early in the century; the first time Maillart used it in one of his bridges was at Salgina. Tavanasa and the other early three-hinged bridges all had soft metal plates between two concrete surfaces, while Stauffacher had cast steel hinges.

The concrete hinges that Maillart began to use at Salgina were controversial because they were difficult to analyze, and the idea of a hinge of solid concrete was at odds with the intuition of many engineers.[22] What Roš objected to was not the idea of the hinge, but rather the amount of reinforcement Maillart provided to connect both the arch with the foundations and the two arch halves together at the crown.[23]

Maillart's response characteristically argued primarily on the basis of other completed works, in this case the Pont Candelier, a railroad bridge of 64 meters' span and 6.4 meters' rise. It was also built with hinges, which had about three times the concrete stresses of Salgina and over twenty-five times the steel stresses. Thus, as he said, "The stresses under full load do not even reach those which, in the Pont Candelier, occur for live load alone. A strong reduction in the steel [at Salgina] would, therefore, be very fully defensible. All the more reason why a strengthening is not necessary."[24] Although he did respond to some of the technical points in substance, the thrust of his argument came from successful experience with completed works. In the next two points, he invoked the twenty-two-year-old Tavanasa as justification for a design for Salgina that had lower stresses. In spite of this disagreement over the hinge strengthening, Maillart did agree with three of Roš's points. Roš's answer was mildly worded, but firm, and apparently all his recommended changes were put into effect. He noted that "the Salginatobel bridge remains, even when considering my proposed supplements, fully one of the boldest reinforced-concrete structures in our country."[25] All the while, Maillart and Roš were in close

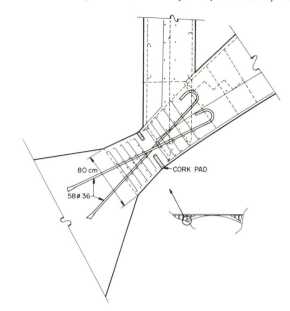

Fig. 8-11. Hinge at springing line, Salginatobel bridge

8-12a

8-12b

8-12c

Figs. 8-12a, 12b and 12c. Construction of the scaffolding for the Salginatobel bridge

personal contact in Zurich, and in his private letters at that time, Maillart, speaking frequently of Roš, never mentioned anything indicating either professional disagreement or personal animosity.[26]

The Salgina scaffolding went forward (Figs. 8-12a and 8-12b) until just before July 23, 1929, when the scaffold designer fell about thirty-five meters (see right side of Fig. 8-12c), surviving miraculously.[27] The Salgina scaffold could not be completed until October, and the concreting would have to wait for the following year.[28] Thus, the bridge moved into a new decade, the last of Maillart's life and the one producing the most innovative yet enduring works. From the design for the Salgina, the district engineer, Lorenz, had grasped something of its significance, and Mirko Roš had immediately seen its boldness; but only Maillart had known what a work it would be once concreted into full form.

Maillart moved through his last decade, the thirties, in poor health, in relative personal isolation, and yet with an increasing mastery of design. Just before the turn of the decade, he created an apartment in his Geneva office, where he went to live. With his daughter gone to Indonesia, all three of his children had left Switzerland. (His elder son Edmond [1902-1962] was in England, and his younger son René [1909-1976] was in France.) His struggles in the twenties had led to some recognition and to close cooperation with several builders, but with the end of the Lorraine, no other large bridge project would come to him, partly because of the depression and partly because he would never again submit a nineteenth-century design for a major bridge competition.

The transition period was effectively over by 1930. Maillart ceased to concern himself any longer with "new developments." He worked primarily from his own inner ideas.[29] During the twenties his reestablishment had meant doing traditional works like Lorraine, but from 1930 on he would work essentially on his own, living alone, riding the train alone between Geneva, Bern, and Zurich.

Two major ideas had taken shape over the previous third of a century in his career as a structural engineer: the deck-stiffened idea with its reinterpretation of the scientific basis for engineering structure, and the three-hinged idea with its reinterpretation of the artistic basis for modern structures. His works would go up in the woods of the Canton of Bern, and occasionally in other cantons. Between 1920 and 1940 he designed thirty-three bridges that were built. There were many other bridges designed but not built.

The Marignier bridge of 1920 appears to have been analyzed and detailed by Maillart following someone else's 1911 plans. The Châtelard aqueduct, while of great interest by itself, does not relate directly to the developments under discussion. It does, however, begin Maillart's study of flat, beam-type bridges which continued with the Nessental footbridge, the bridges at Liesberg and Huttwil, the Gründlischwand, Bern, and Altendorf structures, and the less important Laubegg bridge (see Appendix B). All the other works are in one of the two mainstreams of Maillart's development, either deck-stiffened or three-hinged.

For the three-hinged class, the Salginatobel bridge stimulated Maillart to design eight more that were built: the Rossgraben in 1932 of very similar design to the Salginatobel; the Thur bridge at Felsegg where for the first time he used a broken arch; the Arve bridge at Vessy where he introduced x-shaped cross walls between the deck and the arch; the Lachen overpass; and the Garstatt bridge for which all curves are omitted and the broken arch has a triangular profile. In between he had designed bridges at Innertkirchen, Twann, and Wiler, all without the open profile of the Tavanasa-Salginatobel style.

The deck-stiffened style begun with the Wäggital bridges and the Valtschielbach bridge, developed continuously through the Klosters railway structure of 1930 and on through a series of small works in the Canton of Bern at Adelboden, Frutigen, Schangnau, and Habkern. All show the very thin arch, the much deeper, straight-deck girder, and the slender, vertical cross walls in between. Just as the development of three-hinged arches culminated in the masterpiece at Vessy and the startling angular shapes at Lachen and Garstatt, so the deck-stiffened style found its most stunning expression in its final two works, the curved-plan arch over the Schwandbach of 1933 and the footbridge over the Töss river of 1934.

Despite these final achievements, however, at death Maillart had achieved little recognition in international engineering circles, although he had begun to be noticed as an artist by the avant-garde.[30] It was left to a few dedicated nonengineers, his admirers Sigfried Giedion and Max Bill, and his daughter, Marie-Claire Blumer-Maillart, to persist in collecting, publishing, and exhibiting documents of his work to a larger audience. But as his reputation as an artist grew, his ideals of structural engineering were almost forgotten.

As we move to an assessment of Maillart as an artist, we shall continually keep in mind the inseparability of "the more rational and the more beautiful"; and especially that these two sides to his work incorporate the two aspects of his single ideal. We shall be exploring an art that makes rationality, in all of its scientific force, a principal goal, but that simultaneously retains beauty as an equally important end. For Maillart, this art was the result of aesthetic choices from among many possible rational forms.

9 The Role of Science in Engineering: Force Follows Form

In the twentieth century, engineering came to be thought of by many as an applied science. Engineering design came to be seen as a process that followed logically from scientific analyses developed in theoretical and laboratory research. For the design of large-scale civil structures this picture is, however, at best confusing and at worst wrong.

Physical science—as engineers conceive it—studies a reality existing outside humanity,[1] whereas engineering design centers on human choices of form; the formulas used in design become relative to the visual frame of reference of the designer. For Maillart, the freedom to choose a form set structure off from science, because the search was for a specific form rather than for a general formula. In his ideal of engineering structure, the behavior under loading was so simple and easy to visualize, that the problems of mathematical analysis practically disappeared. There was no need to show the relationship between simple formulas and a general theory because the constructed result would be tested full-scale immediately, and the data made publicly available. The best place to see the distinction between science and engineering is in the deck-stiffened arch bridge designs (listed in Table 9-1).

Table 9-1. **Deck-Stiffened Arch Bridges by Maillart**[a]

Bridge	Date	Span	Rise	Span/Rise	Arch Depth	Deck Depth
Flienglibach	1923	38.7	5.17	7.45	0.25	1.61
Schrähbach	1924	28.8	4.02	7.2	0.18	1.24
Valtschielbach	1925	43.2	5.2	8.3	0.23	1.24
Landquart (RR)[b]	1930	30.0	7.9	3.8	0.26	1.40
Spital	1931	30.0	3.26	9.2	0.24	0.90
Landholz (FB)[b]	1931	26.0	3.40	7.65	0.16	1.20
Hombach	1931	21.0	3.0	7.0	0.16	0.70
Luterstalden	1931	12.5	2.55	4.9	0.16	0.60
Traubach	1932	40.0	5.60	7.15	0.20	1.34
Bohlbach	1932	14.4	2.70	5.3	~0.16	~1.20
Schwandbach	1933	37.4	6.0	6.23	0.20	0.90
Töss (FB)[b]	1934	38.0	3.50	10.84	0.14	0.54

[a] All dates and dimensions taken from documents in PMA. All dimensions in meters.
[b] RR = railroad bridge, FB = foot bridge, all others are roadway bridges.

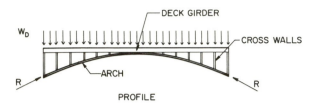

Fig. 9-1. Profile of a deck-stiffened arch

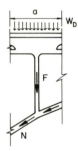

Fig. 9-2. Detail of a loaded deck-stiffened arch showing forces in cross wall and arch

Looking carefully at these bridges, one can see the way in which Maillart's ideas on analysis remained constant, while his ideas on design continuously evolved. In fact, his attitude toward the scientific basis of engineering was fully clear by 1923 in the Flienglibach calculations, and with the Valtschielbach, that attitude found an even simpler expression. Moreover, Maillart attached considerable significance to the Valtschielbach bridge (see Fig. 8-1) as his first major deck-stiffened design.[2] His 1925 computations for it take up just four pages and the entire analysis for the most complicated part of the arch calculations (for asymmetrical loads) was done in less than half a page.[3]

The essence of his method lay in a first assumption that the arch does not bend under live load. More precisely, Maillart assumed that the arch stresses due to live-load bending were so small that they could be ignored. Therefore, he assumed that the stiffening girder carried all the live-load bending. The second assumption took this girder bending to be numerically equal to that which the arch would have had to take were it unstiffened. Under these two assumptions, Maillart then made a structural analysis to determine the forces in both the arch and the girder, and finally, on the basis of that analysis, he computed both the concrete compression stresses in the arch and the reinforcing steel required in the girder.

For his two assumptions to be reasonable, the form of the structure had to be severely restricted, that is, these assumptions followed directly from Maillart's choice of form. Without this form his method of structural analysis would have made no sense at all; but with it, this was the most reasonable way to proceed.

As a description of the manner in which the bridge carries its loads, Maillart's analysis best fits his chosen form. It is sometimes said that the form should follow the forces, but as we shall see here, Maillart's idea was quite different. For him, the forces followed from the choice of form, and the analysis required to define those forces also followed from it.

By *force* we do not mean external forces such as gravity or wind, which the structural engineer calls loads. Force here always means the structural action inside a form; thus, forces define how the structure reacts to, or carries, its loads. Given a set of loads, the form will determine the forces. The proper name for the exercise directed toward determining these forces is *structural analysis*, while the resulting forces and displacements define the *structural behavior*. To see this in practice requires a brief description of the forces in a deck-stiffened arch.

The Structural Analysis of Deck-Stiffened Arches

The stiffened arch system of Figure 9-1 carries uniform loads (dead load, snow, or traffic spread evenly over the span) by axial forces alone. These terminate in foundation reactions R in the direction tangential to the arch axis at its springings. We can imagine the uniform load W_D (kg per meter) on the deck to be carried by vertical axial forces equal to F in each of the vertical cross walls spaced the distance a apart (Fig. 9-2). The arch in turn carries these vertical forces by sloping axial forces N. The sloping arch members carry the vertical force F because they are kinked at their intersection, that is, they have different slopes. Thus, the load on the deck goes right through the cross wall to the arch and thence to the foundations, all without bending the arch.

When the deck is loaded by a live load W_L over only half the arch span, the same behavior cannot occur. We begin by assuming that the load does go right through the deck to the arch, but, as shown in Figure 9-3, without the load on the right side, the left side will bend downward and the right upward. When this happens, the deck must follow because it is tied to the arch by the cross walls. As the right half of the arch pushes upward, the deck will resist by an amount that depends upon its stiffness. If the deck is very thin, it provides almost no resistance and the arch must deflect (Δ) to carry the half load alone (Fig. 9-3a). If the deck is very stiff (Fig. 9-3b), it provides such a great resistance that, in effect, as the arch tries to lift on the right, the deck reacts and produces forces in the vertical members that push the arch back down, thus drastically reducing the deflection and producing nearly uniform forces on the arch just as if the deck were uniformly loaded as before. This time, however, the uniform arch load, in effect, is half the full load W_L and the deck carries up and down loading as shown in Figure 9-3b. The result is a uniformly loaded arch with axial forces only and a nonuniformly loaded deck with bending forces only. This simplified discussion demonstrates Maillart's method of analysis. He computed the bending forces from an arch analysis and then assumed those values in the deck girder. The analysis presented here is precisely that used by Maillart for the Valtschielbach bridge in 1925 and, in principle, is the same as for his 1933 Schwandbach bridge.[4]

Maillart, having developed his method of analysis before 1925, was later free to concentrate his attention on design questions. Although the differences in analysis between the Valtschielbach and the Schwandbach were negligible, the differences

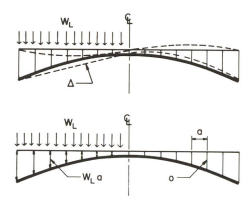

Fig. 9-3a. Structural behavior of unstiffened arches

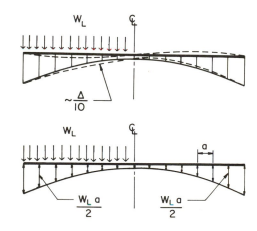

Fig. 9-3b. Structural behavior of stiffened arches

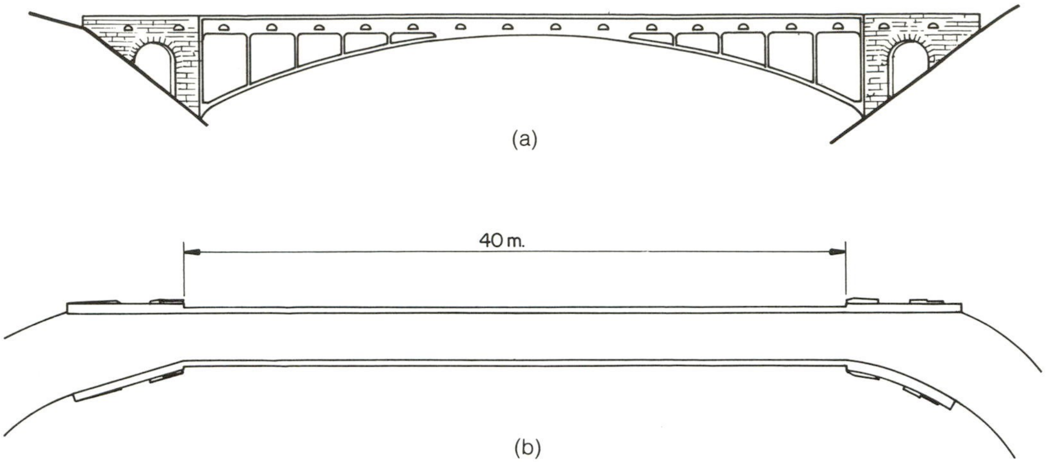

Figs. 9-4a and 4b. (a) Elevation and (b) plan of the Valtschiel-bach bridge, 1925

in design were great. In 1925 Maillart produced a bridge that was technically significant and visually striking; in 1933 he designed one of the two or three most beautiful concrete bridges ever built. It is just his priority of design over analysis that led to this development. The differences show clearly when we contrast the two bridges.

The Structural Design of Deck-Stiffened Arches

By *design* is meant explicitly the giving of form. Frequently in American texts on concrete structures the term design refers to computations that determine both the size of the concrete stresses and the amount of reinforcing steel needed. Here, *design* means the choosing of form, rather than merely the checking of dimensions (which results from computation of stresses) or the determining of embedded reinforcement.

To compare the designs of the Valtschielbach and Schwandbach bridges, we shall consider the following typical bridge features: first, the plan layout; second, the approach structure; third, the parapet and deck; fourth, the connections between arch and deck; and fifth, the arch itself. In each case Maillart in the later bridge introduced substantial changes from his earlier design, and each time these differences marked the unfolding of design ideas freed from concern about analysis.

First, the plan layouts demonstrate clearly this maturing because both bridges are on roadways that make curved turns across a ravine. In 1925, Maillart took the bridge straight across and introduced sharp transition curves at either end of the

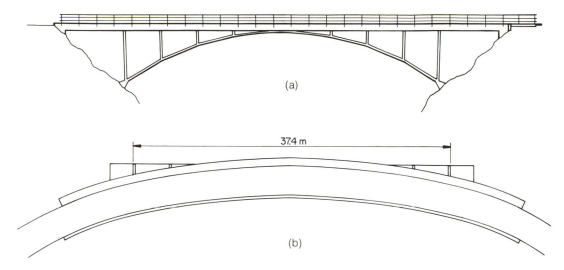

(a)

37.4 m

(b)

Figs. 9-5a and 5b. (a) Elevation and (b) plan of the Schwandbach bridge, 1933

approaches (Fig. 9-4b). In 1933, he formed the structure on an elliptic ground plan, giving the Schwandbach its doubly curved appearance and allowing the roadway to cross the ravine smoothly with no sharp transitions (Fig. 9-5b). In the mind of the analyst, this ground plan curve creates great difficulties because it violates assumptions essential to an analytic formulation.[5] But for the designer, it permitted a new study of form because of the need to connect the horizontally curved deck to the vertically curved arch and still retain a structure whose behavior could be easily described. Maillart saw this connection as one in which the thin cross walls, previously straight at Valtschielbach, would be sloped outward from the deck at Schwandbach to meet the arch at different widths (Fig. 9-6). In this way the arch stayed straight on the outer side while being splayed on the inner side to follow the deck curve. The arch had a wider section at its springings, making it stiffer against horizontal loads. This is commonly done on high bridges, where the combination of vertical gravity loads and horizontal wind loads produce the same type of twisting behavior that the horizontally curved vertical deck loads produce on the Schwandbach.[6] By widening the arch base to retain the essentially planar behavior, Maillart modified the Valtschielbach form to meet the loading created by the smoothly curved plan layout.

The second major difference between these two deck-stiffened arches occurred in the approaches. At Valtschielbach (Fig. 9-4a), Maillart used a heavy masonry structure. Although this may seem curious, it was typical of Maillart to make design changes slowly, and his major bridges prior to 1925 had masonry or solid wall approach structures. It was natural, as he thought about the deck-stiffening form in its early years, that he would continue to use other ideas he had not yet questioned. Even on some of the deck-stiffened bridges built between 1925 and 1932, he still used closed masonry-like concrete walls for the approaches. But on the Schwandbach, the approaches suddenly were open and of the same form as the main span (Fig. 9-5a). Just as the transition to the roadway was smooth and integral, so was the transition from the mainspan to the hills on either side. This transition emphasized the stiffness of the deck girder which, over the mainspan, removed bending from the arch; while over the approach spans it carried all the load directly. A deck girder just deep enough merely to carry the mainspan loads to the cross walls would have been a good deal shallower than the girder needed to carry loads over the much wider approach span. But by using it as a stiffener over the arch, Maillart was able to carry it visually at the same depth over to the abutments. All traces of past forms and of masonry materials were gone.

One other trace of the masonry past, and the third contrast, appeared at Valtschielbach in the parapet, which was a solid wall pierced by semicircular openings of the same Romanesque quality found in the approaches (Fig. 9-4a). Heavy stone parapets were common on pre-twentieth-century masonry bridges, especially in the Graubünden, where such parapets with semicircular openings frequently appeared on stone bridges built in the late nineteenth century.[7] Maillart had taken an old form of parapet and turned it into structure much as, earlier at Zuoz, he had taken from Stauffacher the old form of longitudinal spandrel wall and made it into an integral part of the new bridge structure. And just as the Zuoz wall was an early and only partly developed form, so was the Valtschielbach parapet, as we see immediately on examining the Schwandbach (Fig. 9-5a). Here the parapet itself became merely a light metal railing, while the stiffening member was now a longitudinal deck girder partly above and partly below the roadway surface to which it served as a curb, but not as a parapet. The result was a lighter-looking structure with formwork and reinforcing less complicated than the wall with semicircular openings. Tavanasa did not resemble the Stauffacher from which its form developed, and Schwandbach did not look like the Aarburg bridge of 1911, nor did its parapets resemble those of nineteenth-century stone bridges.

The fourth difference in this eight-year interval arose from the way that the deck girder and the arch met. At Schwandbach the arch, by barely touching the girder, appeared continuous, while at Valtschielbach it merged with the parapet girder over a large part of its middle length.

Fig. 9-6. Schwandbach bridge near Hinterfultigen by Maillart, 1933

Fig. 9-7. Töss River bridge near Wülflingen by Maillart and W. Pfeiffer, 1934

Finally, and almost imperceptibly, the arch itself at Valtschielbach was curved along its underside, whereas at Schwandbach it was polygonal; so that in spite of its highly curved overall appearance, Schwandbach is essentially a structure of straight members.

Maillart had taken the idea of a form, first displayed in the 1923 Flienglibach bridge, and then in the Valtschielbach, and developed its design into the masterpieces of Schwandbach and, a year later, the footbridge over the Töss (Fig. 9-7). In large measure this development was related directly to Maillart's belief in the priority of design over analysis or, more properly, in the subordinate role of analysis as a tool of design. As Maillart himself said near the end of his life, "Only a fully simplified method of analysis is . . . both possible and sufficient. The discerning use of analytical results leads to structures consistently safer than those based on the strict but thoughtless use of a refined and sophisticated method of analysis [*einer mit allen Feinheiten ausgestatteten Rechnungsmethode*]."[8] In engineering theory of the twentieth century, "sophistication" and "refinement" (one might also add "rigor" and "generality") are as overvalued as they were when Maillart wrote in 1909.[9]

Maillart's ideas become even clearer when one compares them closely with those of his contemporaries, who placed such value on the necessity of sophisticated analysis for engineering design. Two major research works, one published by the American Society of Civil Engineers (1935) and the other (1942) directed by a leading Swiss academic, Max Ritter (no relation to Wilhelm Ritter), embodied this belief in sophistication and, significantly, did so while focusing directly on deck-stiffened arches. A third major report by another Swiss academic, Mirko Roš, appeared about the same time (1937) and essentially presented Maillart's viewpoint.

American Arch Studies: 1923-1935

As Maillart's last deck-stiffened bridge was being completed, a committee of the American Society of Civil Engineers published a long report on "Concrete and Reinforced Concrete Arches."[10] Covering research over exactly the same eleven-year period during which Maillart had developed his designs from Flienglibach to Schwandbach, the committee, like Maillart, had studied mainly arch ribs and arches with open spandrels.[11] Since members were well aware of European work,[12] and since Maillart had followed developments in America, it is truly astounding, at first glance, to realize that not only was there no reference by the committee to Maillart's work, but that their ideas had almost nothing in common.

The report drew on nineteen separate studies that had originated mainly in university research projects, including experiments on both concrete and plastic models. The studies centered on single-span arches and on three-span arches on slender piers, both with and without deck girders. The effects of climate, foundation movements, and creep (gradual movement due to compression) were analyzed, with records taken from the behavior of numerous concrete arch bridges in service. The committee claimed to have studied the "design and construction" of reinforced concrete arches; yet the report turned out to be purely analytical.

One conclusion is important enough to be quoted in full.

The following method of analyzing the stresses in a single-span arch bridge with open spandrel and deck is suggested in the design of structures for which deck participation may be important: (a) determine the dead-load stresses analytically on the basis that the dead load is carried by the rib [arch] unrestrained by the deck. (b) determine the live-load stresses by an experimental method using elastic models, considering the structure as a whole and not the rib alone. This is important because of the effect of the deck participation upon the stresses in the deck and columns rather than because of its effect upon the stresses in the rib. (c) determine the shrinkage and temperature stresses by an experimental method using elastic models of the structure as a whole.

The theoretical analysis of an arch with open spandrel and deck is so complicated that few engineers have acquired the facility that justifies them in using it with confidence. Furthermore, the time required for such a theoretical analysis is excessive.[13]

The analysis for deck participation was conceded to be complicated and to take excessive time, but more important, deck participation (b) was presented as a problem in analysis rather than a possibility for design. There was no indication that by making a relatively stiff girder the arch could thereby be made much thinner. In the concrete model tests, the deck and arch were about equally stiff, which led to substantial arch bending, and thus the need for the heavy arch. Indeed (b) seems to imply that the arch stress would not be greatly influenced by the participation of the deck, an implication that may have led designers away from the kind of insight characteristic of Maillart.

The committee also observed that expansion joints in the deck are not recommended and that deck participation does reduce live-load bending at the arch springings. Substantial time was devoted to studying the influence of removing the stiffening effect of the girders, just the effect Maillart used to advantage. The contrast here is significant because it shows the Americans' tendency to use advanced analytic techniques (models in the 1935 report) for handling complex details, without explaining overall behavior.

While we have no record of Maillart's judgment of this report, a contemporary study of Swiss arch construction by Mirko Roš and others seems to represent his viewpoint, developing design principles from the study of many structures rather than prescribing specific designs from physical theory and laboratory tests.[14]

This crucial difference was due to the models tested: full-scale, complete structures in the Swiss report and small-scale, partial structures in the American report. The American report, in its attempt at generality, recommended that analysis be done by the use of elastic models because theoretical analysis was so difficult. Such an idea encourages the designer to put off design decisions until the results of his analysis are available, when in fact, the very choice of model dimension is the most important design decision of all. Elastic models are extremely difficult to use in design because they cannot easily be changed to allow study of different forms. Field studies, as Maillart knew, force the designer to think about structural behavior. A bridge incorporates the creativity of the visual arts and the discipline of physical science. Its visual form, like any artistic conception, can be judged by the aesthetic criteria of our culture. However, once created, a bridge must also obey the laws of nature, and its functional efficiency can be understood only by an analysis of its structural behavior using tools similar to those of physics.

In the next chapter, the consequences of the relationship between engineering and art are explored; but here, to complete the discussion of deck stiffening, we shall further interpret Maillart's ideas about engineering and science by contrasting his work with a third major discussion of deck stiffening, appearing in the form of a publication from the ETH just after his death.

Swiss Arch Studies 1942

An ETH doctoral dissertation under the direction of Professor Max Ritter by a student named Abd-el-Aziz El-Arousy is especially revealing, dealing with the elastic behavior of bridge arches that act together with their roadway decks to carry loads.[15] It represents a powerful academic viewpoint on the relationship between scientific analysis and structural design.

El-Arousy's goal was the further improvement and simplification of the analytic calculation of arches, directed towards practical use of these calculations. He was seeking a general method of analysis that, by use of tables and charts, would permit the analysis of arch and deck forms with all types of relative dimensions.[16]

The work is divided into two parts: one on arches alone and the other on arches built together with decks. In the second part, El-Arousy presented clearly the view of analysis against which Maillart had struggled throughout his career. That viewpoint contained two ideas: one was that generality is the essential basis for applications, and the other was that form follows from forces, that is, that structural design of forms is based upon forces determined from a prior structural analysis. El-Arousy gave as his basic idea the use of analysis to obtain internal forces in each of the two main parts (the arch and the stiffening girder) and from such forces "to dimension these two parts correctly according to these forces."[17] In a correct design, following this view, the relative dimensions fit the relative forces obtained from a general analysis. But Maillart's idea was that the relative forces should fit the relative dimensions. For him, the dimensions set by the designer controlled the forces; whereas for El-Arousy and his teachers, and indeed for many writers on structural design in the twentieth century, it was the forces obtained by the analyst that controlled the dimensions.

The Contrast of Ideals between Analyst and Designer

Maillart considered the arch to carry all the loads that are distributed uniformly along its length, these being the weight of the arch itself, the weight of the entire superstructure, and the weight of a snow load or a full loading of traffic. This assumption is consistent with all general analytic theories, but two others are not. First, he considered the deck to take all the bending arising from nonuniform loads; and second, he determined the amount of the deck bending by computing only the bending of an unstiffened arch, the results of which were simply taken to describe the deck for practical purposes. Moreover, he analyzed that unstiffened arch as if it were hinged at its two supports points, while in practice he designed it to be rigidly built into the foundations. Taken together, these assumptions are outlandish, and yet, in each of his completed bridges full-scale load tests always verified them.

El-Arousy's first discussion of a deck-stiffened bridge demonstrates how an overemphasis on analysis restricts design. He analyzed an arch and deck connected by five vertical members for a concentrated load at midspan.[18] His method was complex because it came from a general theory; and he was thus insecure about its validity because, to get a solution, he had to make a few assumptions. In fact, he noted the importance of laboratory testing in getting verification for the analytic procedures.[19]

Following the initial results, he introduced four more vertical members. Because of the complexity of the analysis, he solved this problem also by an approximate method. From the results El-Arousy then concluded that applications of the approximate method required at least nine verticals. He then commented: "This is a happy result because the greater the number of spans [space between vertical members] the more time-consuming is the use of the exact method, whereas the better are the results of the approximate method."[20]

What this study discovered, and what Maillart had grasped nearly twenty years before, is that there are really two separate theories: one applicable to deck-stiffened arch bridges where the connection between the arch and deck can be considered as complete, and one applicable to such bridges where the connections are so widely spaced that they must be considered as isolated supports. The choice between the two theories depends on the previous choice of the bridge form and not on the demonstration of its accuracy compared to a more exact general theory covering all forms. Furthermore, the specification of nine verticals (or eight spans) has little precise meaning because, again, it depends not on the comparison with general theory, but rather on the category of form. For example, on the 37.4-meter Schwandbach bridge, Maillart used ten verticals, while on his proposal for a 100-meter arch over the Gorges du Trient he used seventeen, and on the 12.5-meter arch of the Luterstalden bridge he used only five.

One has only to imagine a 12.5-meter span with nine verticals (thus spaced about 1.4 meters apart) to realize what a curious design would have resulted from "the strict but thoughtless" use of analytic rules. Although for Maillart's approximate method the relative deviation from the more exact analysis results (such as those of El-Arousy) is much greater in the short-span arch, the absolute values are so small as to render them insignificant for dimensioning. Obviously, Maillart recognized the importance of scale in his forms. But his simple method of analysis fit not just some of his deck-stiffened forms, but all of them, because that analysis merely demonstrated numerically what Maillart wanted to express physically—that the arch should be as thin as possible. All aspects of his dimensioning centered on those goals of lighter scaffolding, minimum materials, maximum integration of all structural members, acceptable behavior under full-scale testing, and the permanence of the completed bridge. The simplicity of his analysis contrasted starkly with the complexity of his rationale, because as an engineer, and not a scientist, Maillart cared most about the creation of a specific bridge rather than the discovery of a general form in nature.

In science, the data is taken from nature with as little human interference as possible. In engineering, the data comes from structures that are the artificial products of human design. The goal of scientific theory lies in explaining the natural world, structured by forces disassociated from humanity, whereas the goal of engineering theory lies in explaining the artificial world built by forces associated with human civilization. Hierarchical theories imply fixed objective relationships among all the data, if we were only clever enough to see them; categorical theories, on the other hand, imply data that cannot be related together in one unit, but need to be grouped carefully before any theory makes sense.

For Maillart, a certain restricted category of structural conditions led to the use of deck stiffening and hence, his method of analysis was not directly connected to any general theory of analysis at all. Indeed, it radically violated all such connections. Maillart himself defined that restricted category in a 1933 letter responding to an inquiry from Milo A. Ketchum Jr., then a graduate student at the University of Illinois. Maillart wrote that "the stiffened polygonal arches are especially suitable for narrow bridges where the parapets can be used as the stiffening girders."[21] Thus his deck-stiffening idea arose out of a specific condition typical of small mountain roads. Using that idea, he designed the relative dimensions so that, as he noted on the Flienglibach bridge calculations, the arch bending would be negligible.

The validity of Maillart's analysis is shown vividly in Figure 9-8. El-Arousy demonstrated with this diagram how the moment in the arch becomes insignificant when the deck girder is relatively stiff. He noted that the simple method had been proven by Max Ritter in a work not yet published by 1942.[22] This method was just that used by Maillart in 1923 for the Flienglibach bridge. When El-Arousy showed how the arch moments diminished radically when the ratio of girder to arch stiffness increases from one to ten, he derived no conclusions from that diminution. His only interest was to show how an approximate method of analysis differs from a more exact method. From a designer's point of view, the most significant engineering conclusion is missing, even though the data are there in the examples chosen for illustrations. The principal reason for this oversight lies in a vision of structures very different from Maillart's. We can make this clear by a simple example, using a scientific analysis.

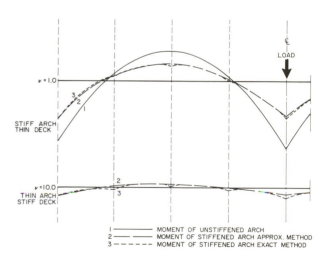

Fig. 9-8. Structural behavior of unstiffened and of stiffened arches (taken from A. El-Arousy, *Mitteilung 13 aus dem Institut für Baustatik*, Zurich, 1942, p. 116)

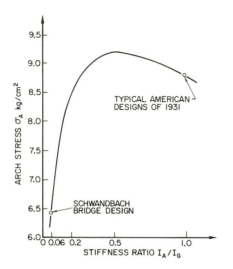

Fig. 9-9. The influence of deck stiffening on arch bending stresses

Scientific Analysis and Engineering Vision

The following question illustrates the use of analysis in the context of design: What is the optimum ratio of arch stiffness I_A to deck girder stiffness I_G needed to produce the lowest stresses in the arch under a live load spread uniformly over half the deck (Fig. 9-3)?

The answer to this question, formulated and solved rigorously, leads to the graph shown in Figure 9-9, in which the arch stress σ_A increases upward and the relative arch stiffness increases toward the right.[23] The question surprisingly turns out to have two answers, neither of which are of any practical use; but the implications are crucial.

The first answer comes from that part of the graph past the ratio of I_A/I_G greater than one-half, that is, the descending right-hand part of the curve on Figure 9-9. This part shows that an increase in arch stiffness (for the same girder stiffness) leads to a *decrease* in arch stresses, and hence, appears to be desirable. On the other hand, a second answer arises if we look to the left-hand part, in which an increase in arch stiffness (again for the same girder stiffness) leads to an *increase* in arch stresses and thus, seems to be undesirable.

Furthermore, neither answer is of direct use because neither gives an optimum, that is, the lowest stress in the arch. To the left, σ_A becomes a minimum only at $h_A = 0$ and to the right σ_A becomes a minimum only where $h_A = \infty$. It is of no practical use to find that the stress is a minimum either when the member itself vanishes or becomes infinitely thick. There are practical limitations to both the minimum and the maximum sizes of arch. But within those practical ranges there are two completely opposite answers to the same question; the answer one gets depends upon the starting point.

If the designer thinks in terms of heavy arches, such as those of stone, then the starting point on Figure 9-9 will be to the right and his tendency will be to view the deck-stiffening as a disadvantage. As the arch becomes stiffer, relative to the deck, the arch stresses decrease so that the heavier the arch the better. By contrast, when the designer envisages light arches, the scientific analysis shows deck-stiffening to be an advantage and the lighter the arch the better. The crucial difference is in vision.

Let us return to the distinction between engineering and science. The scientific idea (in the sense of engineering sciences) of describing form is hierarchical, leading from theories which are general or exact to ones which are specific or approximate; while the engineering idea (engineering in the sense of civil-works structures) of describing form is categorical, with each form having an appropriate and simple theory. Engineering form is the product of human imagination, not of natural forces. There is no such thing in nature as a parabolic arch, let alone a deck-stiffened arch. There are suggestive forms in nature that demonstrate ideas of stiffening and of load bearing, but they appear to play no direct role in the imagination of the structural engineer.[24]

The engineer cannot choose form as freely as a sculptor, but he is not restricted to the discovery of preexisting forms as the scientist is. The engineer invents form, and Maillart's career shows that such invention has both a visual and a rational basis. When either is denied, then engineering design ceases. For Maillart, the dimensions were not to be determined by the calculations, and even the calculation results could be changed because a designer rather than an analyst was at work.[25] Analysis and calculations were servants of design.

Fig. 10-1. Salginatobel bridge by Maillart, 1930

10 The Role of Art in Engineering: Structure as Art

The deck-stiffened designs define the relationship of Maillart's ideas on structural engineering to applied science. The three-hinged arches permit exploration of his ideas on structural engineering considered as a part of the visual arts.

Maillart had a self-assurance about form that arose from familiarity with analysis, but that had been conditioned heavily by empirical evidence from construction experience and from full-scale testing. This separation of structural engineering design from scientific analysis has led many observers to believe that engineering designs can be categorized with the visual arts; that structural engineers like Maillart are artists whose works are to be considered somewhere between sculpture and architecture. They have imagined that the essence of the artist as engineer lies in his need to deny the analytic side of structure. Even Sigfried Giedion called the deck-stiffened arch "the most daring of his constructions, the least verifiable by calculation and the most opposed to the ruling taste."[1] But Maillart was an *engineer*-artist; all his bridges are directly verifiable by calculation, if by calculation is meant the numerical prediction of full-scale physical behavior. Contemporary structural analysts had objected to the difficulty of analyzing his designs by generalized scientific methods. As the detailed test reports of Roš show, Maillart's bridges were as verifiable as many others; and, considering the simplicity of his analysis, his may have been easier to verify than those of his contemporaries.

Maillart's fundamental idea was that structure should be liberated from mathematical analysis; but, at the same time it should be disciplined by the results of physical testing and visual observation. Both sides of this basic idea emphasized the visual result rather than the numerical process. But the freedom to visualize form received its license from the record of tests and observations.

Maillart was a keen observer of completed structures, and as the examples of Zuoz and Aarburg show, his leading ideas arose from thinking deeply about how to improve full-scale behavior. But in that process of thought, he always felt he had a choice among many possible and technically reasonable design modifications. His criterion was appearance, as the example of deck stiffening makes clear. To avoid cracking, the choice of modification was either to cut joints in the deck girders (build in the cracks), or to design the arch to take all the load and make the deck flexible enough so that it was subjected to less force. It would be extremely difficult to establish any significant economic difference between these options. Therefore, Maillart's choice was not primarily determined by economics, but rather, within a set of equally inexpensive possibilities, he selected the one that appealed to him visually. Our judgment of his works in the context of those of his contemporaries can thus be aesthetic. Maillart's three-hinged arches are especially good structures to study as works of art because of their variety and the extent of his writing about them.

Attempts to define art, let alone engineering as an art, can lead directly into philosophical questions beyond the scope of this book. My objective is to recapture what seems to have been an earlier viewpoint on building held by builders like Maillart early in the twentieth century and to discuss this in the light of some ideas that arose within the modern movement during the last twenty years of Maillart's life.

The earlier viewpoint was that of the builder discovering for himself the forms possible, the properties of his materials, and the means of putting them together. This implied someone who was, as the historian would say, working with primary sources rather than someone like a teacher who must necessarily use secondary works. It is commonplace to observe that one generation's scholarship becomes the succeeding generation's textbooks. When a new kind of building appears, it is difficult for textbooks to do justice to it until the pioneering works have been built. Unfortunately as writers try to formalize and generalize the results of such completed works, they often neglect the special conditions under which most pioneering works such as Maillart's bridges have arisen.

First of all, these bridges were public structures, and so their designer was forced to set his ideas within a public landscape. This contrasts with what Reyner Banham identified as the central focus of designers in the modern movement, the private home. One of the two works singled out in his last chapter in *Theory and Design in the First Machine Age* was designed in the same year as the Salginatobel bridge (1928) but it represented an entirely different vision, far removed from the public landscape. "Les Heures Claires," a house for the Savoye family at Poissy-sur-Seine, designed by Le Corbusier and completed in 1930, does not reflect its surroundings; it was essentially centered on private living rather than civic utility.[2]

The second special condition for Maillart's work was his exclusive commitment to concrete. While he experimented continually with both structural and aesthetic design, he never doubted that his one medium was reinforced concrete. He believed that concrete was the building material of the twentieth century and, starting with his article on Hennebique in 1901 and his speech of 1904 in Basel, he spoke and wrote in that vein until the end of his life. Moreover, he studied and wrote widely about the properties of reinforced concrete as a physical material, not just as a means to visual form. These papers, considered completely apart from both his designs and his writings about design, would qualify Maillart as an important technical writer of his period. For example, his widely quoted article, "Current Questions in Reinforced Concrete Structures" (1938), is really a detailed technical discussion of the analysis of reinforced concrete cross sections.[3] The relatively few comments on form, the only part of this article included in Max Bill's book on Maillart, were merely an introduction to his main topic.

Having found his material, Maillart gradually developed a distinctive style: light, straight, and exposed, with few curves, and a minimum of decoration. It required a balance among many conflicting objectives. For example, Maillart strove for minimum use of materials and for minimum cost, but field labor costs can be high when very thin sections are designed, because forming and casting require more precision. Along with minimum curves, he sought maximum expression of the overall form; and to minimize applied decoration, he tried to achieve detailed shapes and textures within the structural form itself.

These conflicting goals are never more clearly resolved than in the Salginatobel. This structure is probably better documented than any other concrete bridge. Publications about it include: all the calculations, with a detailed commentary;[4] an article discussing its place in Maillart's career, with a technical explication of its form;[5] a full description of its construction;[6] a complete discussion of its load testing;[7] articles on the origin of its form;[8] and writings on its appearance and its symbolic value.[9] While personal judgment as to its artistic merits is ultimately crucial to the present argument, it is possible to establish the technical excellence of the Salginatobel bridge factually. We can describe Maillart's objectives through close analysis of this bridge, using the evidence of material quantities, costs, dimensions, construction procedure, and the details of surface and shape.

Maillart resolved the conflict between minimum materials and minimum cost by designing a form in which the construction procedure permitted very light scaffolding. The bottom curved slab became not only an integral part of the final hollow-box arch, but also a part of the construction support for the vertical walls and horizontal roadway. When these latter elements were cast, they needed full support until sufficiently hardened to carry their loads; during that critical hardening period (about the first twenty-eight days), the lower slab carried the superstructure and the light scaffolding merely braced that thin arch temporarily. In this way, the scaffold needed to carry only the thin arch, which was slightly less than 30 percent of the total concrete

Fig. 10-2a. Visp River bridge near Stalden by Sarrasin, 1930

dead weight.[10] Thus, by making the lower slab as light as possible, Maillart significantly reduced the scaffolding, which is a major part of construction cost; he had used the construction method described above as early as 1901.

For his objective of minimum curves with maximum expression of the overall form, Maillart removed all arcading from the approaches and used straight elements for all the approach frames, vertical cross walls, and deck spans between. Even the main arch was formed entirely of straight segments, although its overall profile is curved approximately to a parabola along the underside.[11] Compare the profile of the Salgina to that of the Stalden bridge of Alexandre Sarrasin completed at the same time (Figs. 10-1 and 10-2a). Without detracting from the impressive Sarrasin bridge, a comparison shows clearly how much more striking is Maillart's form in which the structurally unnecessary, small arched deck girders or arcading found in the Stalden bridge are gone, and a single arch form predominates. The Salgina profile thus depends strongly on its simplicity of form achieved by eliminating curves, whereas the Stalden imitates in part its stone neighbor built in the nineteenth century (Fig. 10-2b).

Finally, by removing the usual masonry approaches, Maillart eliminated all traces of traditional decoration. He left exposed the uncovered texture of concrete and the structural shapes of the crown hinge, the bottom arch flange, and the slender cross walls. He also carefully related the approach spans to the roadway spans over the arch, while in the Stalden work the approach spans are heavier and columns over the arch abutments substantially wider than the very thin columns above the arch. The Stalden work is extremely light and skeletal but the designer did not integrate all parts together as Maillart did with the Salgina.[12] In this exposure of detailed shapes and of concrete texture, Maillart was genuinely prophetic, even though often misunderstood.

The curator of the Museum of Modern Art's Department of Architecture, Elizabeth Mock, writing in 1949, grasped and promoted Maillart's ideals: calculation merely as a means, concrete as a new material with the unique property of plasticity, and structures designed for economy leading to a continuity which uses their "ultimate strengths." But she thought his bridges "prophesy a future in which welded steel and plastic-bonded plywood, like reinforced concrete, will be molded into thin shells, stiffened by bending. Abandoning line for curves, and two dimensions for three, a bridge will become, more than ever before, a single splendid gesture dedicated to the conquest of space."[13] She was not alone in imagining that the plasticity of concrete implies a great new freedom in form making, with her references to sur-

Fig. 10-2b. Visp River bridge near Stalden, stone, nineteenth century

faces, shells, curves, three dimensions, and splendid gestures. But as those who pay for gestures have learned to their dismay, such is not the case. Concrete's plasticity (its fluid state when cast) introduces the economic straitjacket of form and scaffold. It is just its fluid character that limits its form. The cost of form and scaffold on even standard concrete structures is normally from 35 to 60 percent of the total structural cost.[14] Since Maillart, as a builder, knew this central fact, all his work reflects it. His arch-bridge forms began with the construction of the bottom curved slab, made as thin as possible so that the scaffold could be made with maximum economy. The contemporary Swiss bridge designer, Christian Menn, has described how this idea for economical construction was central to Maillart's practice and how such ideas are basic to the understanding not only of Maillart's works but of post-World War II design as well.[15]

As for curves, Maillart carefully restricted them to very elementary types. In the Schwandbach, for example, the curves of the arch (vertical) and deck (horizontal) are carefully separated so that the forming will be as simple as possible. In fact, the Schwandbach arch is barely curved; it is essentially polygonal, changing its angle at the junction of each vertical spandrel wall. Maillart called these deck-stiffened arch bridges *Stabbogen*, which means straight-chorded arches. Even the curved stiffen-

Fig. 10-3. Simme River bridge at Garstatt by Maillart, 1940

ing girders were formed entirely by straight vertical form boards. Moreover, close study of Maillart's other bridges reveals an avoidance of curves. The deck-stiffened arches, including Schwandbach, are polygonal; the three-hinged arches progress toward the totally straight lines of Garstatt (1940) and away from any curved surfaces (Fig. 10-3). Indeed, as we shall see, he actually criticized the Salginatobel for its curved profile. Moreover, in some of his last bridges, as well as in late unbuilt projects, he began more and more to use straight girders rather than arches. As is clearly demonstrated in his designs for bridges during the 1930's, Maillart gradually abandoned curves for right angles, surfaces for lines, shells (curved slabs) for skeletons (frameworks), and in so doing created splendid gestures in profile and section.

What Maillart prophesied has already come to pass in the works of his followers in Switzerland and elsewhere. Thus when Menn began to design bridges in the late 1950's he could find no more economical solutions to canyon bridging than Maillart's. As Menn said about his Letziwald bridge (Fig. 10-4), "comparisons have shown that the principle of construction used for the Tavanasa bridge was still, 50 years later, the most advantageous one." Again for the Cröt bridge (Fig. 10-5), Menn described the same direct influence, this time for a deck-stiffened design. Finally in getting beyond Maillart's forms to approach his principles, Menn in his mature work (Figs. 10-6 and 10-7) has shown that by beginning with construction, scaffold, and formwork, new and stunning bridges are possible with the post-Maillart development of prestressing. It was just this approach that begins with existing, well-conceived forms and goes beyond them to find the underlying principles that explains best not just Menn's development, but also that of Maillart as he himself described it in 1935.

Prompted by a 1934 article in *Le Génie civil*[16] concerning a French three-hinged arch bridge designed by Charles Rabut and bearing a superficial visual similarity to his own works, Maillart wrote a short paper on his own development of this form. "If we want to get the best out of reinforced concrete in bridge construction," he wrote, "we inevitably arrive at forms that are often quite different from those masonry forms we are accustomed to, and for that very reason prone to imitate."[17] The "best" here should be understood as the ideal of minimum materials and minimum cost which "inevitably" led Maillart to new forms. In what followed, Maillart demonstrated how he himself imitated these masonry forms and only gradually arrived at the best use of reinforced concrete.

Fig. 10-4. Averserrhein River bridge near Letziwald by Christian Menn, 1960

> The difficulty of getting these new and unprecedented forms accepted, to say nothing of making them satisfy oneself, has impelled the engineer—and still more the architect called in to collaborate with him—to try and find ways of compromising between traditional and untraditional designs. This raises the question whether such a tendency is justifiable, and whether it would not be better to confine ourselves to forms deliberately based on purely structural principles.[18]

Here, the "best" were "forms deliberately based on purely structural principles," and moreover, a crucial factor was that they had to "satisfy oneself." The implication was of an external and an internal conflict; the former between the engineer's form and the authorities' tastes, and the latter between the results of the engineer's own reflection and his "purely structural" results.[19] Maillart here recognized near the end of his life the battle that he had waged within himself since at least 1901 when Zuoz appeared. For as he stated later on in the article, "As may be seen in Fig. 1 [Fig. 4-1 above], a bridge over the River Thur at Billwil which I built in 1903, this type of construction can be made to give an appearance that differs very slightly from conventional forms. . . . But in retaining a definitely parabolic arch and solid spandrels

Fig. 10-5. Averserrhein River bridge near Cröt by Christian Menn, 1960

Fig. 10-6. Aare River bridge at Felsenau by Christian Menn, 1974

Fig. 10-7. Model of the Ganter bridge near the Simplon pass by Christian Menn. Completion ca. 1980.

certain structural drawbacks have to be faced, which become accentuated with increasing breadths of span."[20] The traditional appearance raised difficulties for the novel structure. In describing the principal difficulty, he concluded that one was "far from achieving the ideal of a full utilization of the material employed."

This ideal he reached more nearly at Tavanasa, where "the realization that, as a link between the arch and the platform, solid spandrels serve no useful purpose except in the middle of the bridge, and that close to the abutments they exert a useless dead weight which is positively a potential danger [*préjudiciable*], had led to the practice of slotting triangular cavities out of them."[21] It may be unclear to the nonengineer just how radical in content this rather matter-of-fact statement of Maillart's was. For he essentially admitted here that his early bridges at Zuoz and Billwil were not only visually imitative, but structurally flawed! I can think of no other twentieth-century engineer who wrote in a technical journal that several of his completed works were potentially faulty.

These judgments made public the results of a long and vigorous reflection on his own works and gave clear evidence of the internal struggle facing an engineer who tried to create works of art. As he emphasized, the Salgina form had its origin in the 1900 structural idea: "The whole bridge [platform, spandrels, and curved slab], and not merely the principal part of it, forms the arch." But this idea had not been sufficient for Maillart, because there were problems with the spandrel walls that he had later overcome both at Tavanasa and Salgina. After criticizing the French bridge by Rabut for imitating older forms, Maillart posed the question: "It need not be disputed that in giving his arch as constant a thickness as possible, the designer tried to approximate conventional forms. But was this necessary? And does the more structurally honest design of the Salgina bridge evince a marked inferiority in an aesthetic sense?" Instead of answering with a resounding "no" as we might expect, Maillart proceeded to criticize even the Salgina: "But even the latter cannot lay claim to complete sincerity of form. Indeed, if both constant and shifting weights are taken into consideration, the extreme curves of the pressures exerted form two lenticular surfaces whose lower contours meet at an acute angle. It was only in the bridge at Felsegg [Fig. 10-8] built as recently as 1933, that I had the chance of realizing a truly logical form."

This logical lens-like form arose from the need to gain greater depth at the quarterspan, especially when the moving traffic or live loading ("shifting weights") is a larger fraction of the total loading. Maillart continued: "In this case the river crossing occurred on a highway built to carry the exceptional loads of heavy main road traffic.

Fig. 10-8. Thur River bridge near Felsegg by Maillart, 1933

There was, therefore, every inducement to employ the adopted system of construction to the extreme limits of its strength, and to be guided solely by structural considerations in the choice of it—hence the decision to use a pointed form of arch."[22] At Salgina the design total live load is only 11 percent of the total load, whereas at Felsegg it is 21 percent. At Salgina the load test consisted of trucks weighing a total of 10 tons, whereas at Felsegg they weighed a total of 25.8 tons per arch.[23] Moreover, the ratio of span to depth at Salgina is 6.9 compared to 8.6 at Felsegg, which means that the stresses even for the same loading would be higher in the later work. Thus, Felsegg represents, as Maillart noted, a much more heavily loaded structure. This new problem, a bridge on a main road, caused Maillart to think again about his Salgina solution and to see in it both the justification for a similar design at Felsegg and the reasons for several significant design changes. His self-criticism of Salgina needs to be seen more as justification for Felsegg than as regret for Salgina. Just as the specific Salgina constraints (longer span, deeper ravine, more lengthy approaches and even, it seems, a more restrictive budget) caused Maillart to develop a form different from Tavanasa, so local constraints at Felsegg again evoked a form substantially modified from his previous work.[24]

The pointed-arch or ogival shape that Maillart gave to the lower side of the Felsegg arch was a result of his ceaseless search for a simpler and less expensive form. In none of his later works did he return to the rounded curves of Salgina. Moving from the overall form to the detailed shapes, Maillart's article ended with a brief discussion of the approaches to the Felsegg bridge:

> Compared with the powerful simplicity of the wide (236 ft.) span of this Felsegg bridge, the type of support usually adopted for approaches—a series of vertical columns cross-braced to each other—seems a rather paltry device. Inclined two-legged buttresses, suitably reinforced on top, were substituted for these, because they assure good lateral stability and reduce the number of separate foundation-points required.[25]

Fig. 10-9. Approach span supports for the Felsegg bridge

Having solved the problem of the correct form for the long-span arch, Maillart began in the Felsegg work to study the cross framing. His primary explanation here was visual: the "powerful simplicity" of the overall form demanded something more than a "paltry device" for the detailed shape of the cross walls. Such a paltry device he had himself used at Aarburg in 1912. Even at Salgina there was nothing special about his cross walls; there they could be called vertical columns cross-braced by thin walls.

Thus, his new ogival arch form suggested a new design for the approaches as well. Maillart responded with columns connected across their tops by a thin beam. Here, because of the wide bridge (6.5 meters forming two lanes, rather than 3.5 meters forming the single lane of Salgina), he essentially removed the central part of the cross walls, leaving only enough to provide "lateral stability." They splay outward at the base like the end walls at Salgina (Fig. 10-9) except that only the outer rim is left at Felsegg, giving an extraordinarily light contrast with the strong broken arch of the main span. Finally, he noted that there need be only two column footings when the columns are built as a single frame instead of as four single columns coming down from each of the four top ribs. It is the order of Maillart's argument in this last example that best exemplified his thinking: first simplicity, then stability and economy.

Maillart began with aesthetics and then, with that basic consideration in mind, looked for the best structure, where "best" meant minimum materials, minimum cost, and minimum applied decoration. This bridge art was, therefore, vision disciplined by technique; and more specifically, a vision of the public landscape formed by

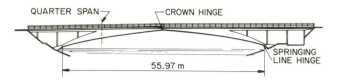

Fig. 10-10. Arve River bridge at Vessy by Maillart, 1936

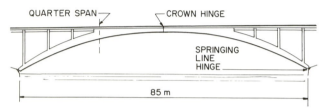

Fig. 10-11. Danube River bridge at Leipheim by E. Mörsch and P. Bonatz, 1937

economic constraints on public structures. It is a difficult art, with the artist continually struggling to control his elements in the face of public opinion, codes, budgets, and politics. We can get deeper into his ideas by looking at one last article written in late 1938.

On September 23, 1938, a detailed article appeared in the journal *Die Bautechnik*[26] describing the eighteen designs submitted for a bridge over the Danube near the town of Leipheim between Stuttgart and Munich. Since none was judged aesthetically satisfactory, the Reich highway office itself made a new design that closely followed a project previously submitted by the firm of Wayss and Freytag in Stuttgart. According to the authors of the article, "in the effort to find the lightest possible structure, this firm has started from the ideas that come from the Swiss bridge engineer Maillart. . . ." The design clearly originated in the three-hinged arch form of the Salgina, and the German authorities understood its structural significance. By removing the vertical wall over the outer quarters of the span, considerable weight was saved without any loss of strength, producing "an extraordinarily light arch rib, in which the material of every part is put to best use."[27]

Not only did the Germans fully credit Maillart for the idea, but they echoed faithfully Maillart's own ideals about lightness, correctness, and the best use of materials. But while agreeing in principle that a light-looking bridge suited the wooded landscape at Leipheim, the authorities nevertheless decided to retain two of the most respected bridge consultants in Germany at that time, Emil Mörsch and Paul Bonatz, instead. Apparently Mörsch examined the structural integrity while Bonatz developed the aesthetic ideas. The difference between a Maillart design (Fig. 10-10) and the Bonatz forms (Fig. 10-11) shows clearly in the cross sections where the lightness of the former (Fig. 10-12) contrasts drastically with the heavy wall-like forms that Bonatz preferred (Fig. 10-13). The *Die Bautechnik* article said of the final result: "One perceives how the entire image of the bridge, through the skeletal structure [removal of longitudinal walls near the abutments], is largely opened up, how it attains by its clean lines a harmony of all of its structural parts, how its elegant span is revealed through the soaring of its light arch ribs, and how the carefully chosen dimensioning of the river piers produces a quietly balanced interaction with the rest of the bridge."[28]

With words like "harmony," "soaring," and "quietly balanced," this critique reveals little of the engineering basis for the form. By contrast, Maillart's critique, while also positive, has a radically different tone, that of a critic who focuses on the engineering meaning fully:

It is gratifying to be able to state that through this work there has been a departure from worn-out paths. Instead of taking over the forms from masonry in which arch and deck are designed to behave structurally as independent elements (without being able to calculate the often damaging effect of interaction), the design here has been worked out successfully to make better use of the nature of reinforced concrete by an appropriate fusion of deck and arch.[29]

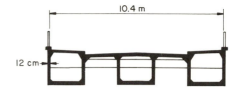

Fig. 10-12. Quarterspan cross section of the Arve bridge

He expressed here, near the end of his life, his pleasure in seeing the German implementation of ideas that he had introduced over thirty years before in Switzerland at the very start of his independent design career. But this pleasure was not unmixed because he saw clearly the differences between the German result and his own views. Even though the scientific idea of deck-arch efficiency was his, the social idea of monumentality was not. At Leipheim, his idea of efficient use of materials had been seriously compromised by the "study of form in consultation with Prof. Bonatz."

After describing the Leipheim bridge over the Danube, Maillart gave a parallel description of his own Arve bridge at Vessy, including its cost of 76,000 francs or about 100 francs per square meter of ground plan, which he contrasted with usual costs of 450 to 700 francs for this type of bridge in Switzerland.[30] By contrast, the German account, although praising the Danube bridge as light, did not give its weight or its cost. Maillart continued with a quantitative comparison of Arve and Danube, which still addressed qualitative questions of appearance and appropriate form, and of the cultural significance of form in public structures of any type.

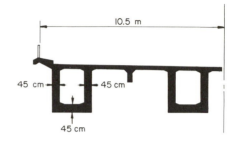

Fig. 10-13. Quarterspan cross section of the Danube bridge

Maillart's primary concerns were efficiency of materials, safety of the entire system, and endurance against the environment. But each of these measurable qualities had to meet a dual requirement, one part physical and the other financial: efficiency and cost, safety and cost, and endurance and cost. In seeking to define a form that in all its details satisfied each pair of goals at once, the designer expressed his ideal, created a symbol of how things should look. As Maillart analyzed these Arve and Danube bridges, he proceeded from the physical question to that of cost, and finally to the visible results.

For the first comparison and the most explicitly quantitative, that of efficiency, Maillart used two measures: one the "boldness ratio" and the other the amount of concrete. The boldness ratio expresses the flatness of the arch, that is, the ratio of span squared over rise. The flatter the arch, the smaller the rise and hence the bolder the design. Similar arches will have higher internal forces corresponding to higher boldness ratios, but flatter arches being lower over the water, will require smaller approach structures. In spite of the higher boldness ratio in the Arve bridge, it required far less material: 0.48 cubic meters per square meter of roadway as opposed to 0.80 in the Danube bridge. According to Maillart, "with about the same concrete stresses, the Arve bridge requires very much smaller dimensions. The only explanation for this lies in the favorable use of the strongly deepened quarterspan section."

Thus, the pointed or Gothic-type arch led to substantial material savings even without increased stresses because "one-sided traffic load creates no need for large expenditure of materials. The dead load is thus of controlling influence. . . ." Even if his flat, pointed arch form was very much more efficient, however, "it will presumably be argued that the Danube bridge is safer, thanks to its larger dimensions." It is true that normally, when the same load is carried by two structures, the heavier one will have a greater overload capacity. But, as Maillart observed, "for shallow arches, the degree of safety depends not mainly on the structure, but rather on the supports. . . . Strong foundations and not large cross sections are, in the first instance, decisive for safety" and, therefore, the heavier Danube bridge was not as safe as the Arve. Maillart then gave evidence of negligible foundation movement, as measured during the Arve bridge's full-scale load test. Such evidence was lacking from the long article on the Danube bridge; the authors mentioned neither load tests nor costs. Maillart implied that the heavier German bridge would be more prone to foundation movements and hence less safe.

After considering efficiency and safety, Maillart turned to endurance by comparing the German attitude toward concrete with his own. "The difference between the concepts may be strikingly expressed in that for the Danube bridge concrete has been treated as, so to speak, something in need of indulgence, to be kept far from the slightest air current, whereas the concrete in the Arve bridge is considered as a strong and healthy substance which in youth or in later years some overexertion and even some roughness will not harm." One characteristic of this difference appears in the hinges, which are of expensive steel in the Danube bridge and of relatively cheap concrete in the Arve bridge. In the latter, the concrete in the hinge is highly overstressed locally, but as Freyssinet and others had shown years before, these stresses do not impair the structure at all.

Growing out of these comparisons, Maillart formed a judgment of the social significance of these two works. "Without regard for anything other than keeping within the available budget," he wrote, "the Arve bridge would be found to be a useful object of the purest type. The Danube bridge by contrast has the character of a monument to the German *Reichsbahn*; it aims to be not a pure useful object, but rather a representative landmark of the German art of building." By starting with the idea of making a symbol first and a bridge second, the Germans had created an inferior bridge. He continued dryly that "the beauty here has been chosen to conform to the type of structure advocated by the collaborating architect Professor P. Bonatz; the resulting structure has thus been made to fit with the aesthetic criterion."[31] Maillart's objections to Bonatz's bridge forms were similar to those leveled against the heavy forms that Albert Speer proposed for the rebuilding of Berlin. Indeed, as George Collins has concluded,[32] Maillart's criticism was cultural and not just structural, and was aimed at the Nazi regime at just the time of its ascendancy.

It may seem that Maillart was largely reacting to the criterion of monumentality and not presenting any criteria of his own. Nowhere did he speak of forms being in harmony either with the surrounding landscape or with the particular traditions of the locality. Examples of self-conscious writing such as we have from some twentieth-century artists are lacking here. However, Maillart was expressing a view of right relationships, even though not in the traditional language of architecture. We cannot expect every artist to set out his aesthetic theory while working to create a new style. Maillart left as complete a record as we have any right to expect.

In his comparative study of the Danube and Arve bridges, Maillart detailed what he opposed. In his works, especially those from Salgina to Arve and beyond, he demonstrated what he believed in: the pure useful object where efficiency, safety, and endurance result from an evaluation and basic understanding of structures already built, tested, and in service. The best work expresses to the fullest the designer's own personal vision of form, and at the same time meets society's constraints on cost (Fig. 10-14).

Fig. 10-14. Arve River bridge at Vessy by Maillart, 1936

Appendix A The Writings of Robert Maillart

Schweizerische Bauzeitung is abbreviated *SBZ*

1901 "Das Hennebique-System und seine Anwendungen," *SBZ* 37, no. 21:225-226.

1901 "Armierte Betonbauten," *Schweizerischer Bau- und Ingenieur-Kalender*, section III : 108-110.

1904 "Neuere Anwendungen des Eisenbetons," *Protokoll der ordentlichen Generalversammlung, am 16 und 17. September 1904*, Basel, Verein schweizerischer Zement- Kalk- und Gypsfabrikanten, Appendix III, 16-23.

1907 "Versuche über die Schubwirkungen bei Eisenbetonträgern," *SBZ*, 49, no. 16: 198-202.

1907 "Belastungsprobe eines Eisenbetonkanals," *SBZ* 50, no. 10:125-127.

1909 "Die Sicherheit der Eisenbetonbauten," *SBZ* 53, no. 9:119-120.

1909 "Armierte Betonbauten," *Schweizerischer Ingenieur-Kalender*, Part 1:268-270.

1912 "Zur Berechnung der Deckenkonstruktionen," *SBZ* 59, no. 22:295-299.

1918 "Die Grundwasser-Vorkommnisse in der Schweiz," *SBZ* 72, no. 5:40-42.

1921 "Zur Frage der Biegung," *SBZ* 77, no. 18:195-197.

1921 "Bemerkungen zur Frage der Biegung," *SBZ* 78, no. 2:18-19.

1921 "Le béton armé à la galerie du Ritom," *Bulletin technique de la suisse romande* 47, no. 17:198-201.

1922 "Ueber Drehung und Biegung," *SBZ* 79, no. 20:254-257.

1922 "De la construction de galeries sous pression intérieure," *Bulletin technique de la suisse romande* 48, nos. 22, 23, 25:256-260, 271-274, 290-293; 1923, 49, nos. 4, 5:41-45, 53-58.

1923 "Ueber Gebirgsdruck," *SBZ* 81, no. 14:168-171.

1923 "Zum Vernietungs-Problem," *SBZ* 82, no. 4:43-45.

1923 "Dispositif de sécurité pour la circulation sur les ponts," *Schweizerische Zeitschrift für Strassenwesen* 9, no. 5:70-71.

1923 "Les autoroutes en Italie," *Schweizerische Zeitschrift für Strassenwesen*, 9, no. 8:109-112.

1924 "Der Schubmittelpunkt," *SBZ* 83, no. 10:109-111.

1924 "Zur Frage des Schubmittelpunktes," *SBZ* 83, no. 15:176-177.

1924 "Zur Frage des Schubmittelpunktes," *SBZ* 83, no. 22:261-262.

1924 "Le centre de glissement," *Bulletin technique de la suisse romande* 50, no. 13:158-162.

1925 "Questions relatives à l'exportation d'énergie électrique et à la mise en valeur de nos forces hydrauliques," *Bulletin technique de la suisse romande* 51, no. 4:41-45.

1925 "Die Brücke in Villeneuve-sur-Lot, nebst Betrachtungen zum Gewölbebau," *SBZ* 85, no. 12:151-154 and no. 13:169-170.

1926 Eine schweizerische Ausführungsform der unterzuglosen Decke-Pilzdecke, Schweizerische ingenieurbauten in Theorie und Praxis, International Kongress für Brückenbau und Hochbau, Zurich, 1926, 21 pages.

1926 "Zur Entwicklung der unterzuglosen Decke in der Schweiz und in Amerika," *SBZ* 87, no. 21:263-265.

1926 "Le centre de glissement," correspondance, *Le Génie civil* 89, no. 14:284.

1928 "Druckbeanspruchung Bei Biegung" First International Congress for Testing Materials, Amsterdam, 1927, Vol II, pp. 13-17.

1928 "Gewölbe-Staumauern mit abgestuften Druckhöhen," *SBZ* 91, no. 15:183-185.

1928 "Die Wahl der Gewölbestärke bei Bogenstaumauern," *SBZ* 92, no. 5:55-56.

1930 "Note sur les ponts voûtés en Suisse," *Premier congrès international du béton et du béton armé*, Liège, September, 7 pages.

1930 "Masse oder Qualität im Betonbau," *Beitrag zur Denkschrift anlässlich des 50-Jährigen Bestehens der Eidg. Materialprüfungsantalt an der Eidg. Technischen Hochschule*, Zurich. Reprinted in 1931, *SBZ* 98, no. 12:149-150.

1931 "Der Brand eines Fabrikgebäudes der Gummifabrik 'Prowodnik' in Riga," *Beton und Eisen* 30, no. 11:206-207.

1931 "Die Lorraine-Brücke über die Aare in Bern," *SBZ* 97, nos. 1, 2, 3, 5:1-3, 17-20, 23-26, 47-49.

1931 "Die Erhaltung des schiefen Turmes in St. Moritz," *SBZ* 98, no. 2:29-31.

1931   "Leichte Eisenbeton-Brücken in der Schweiz," *Der Bauingenieur* 12, no. 10:165-171.

1932   "Zum Entwurf der neuen schweizerischen Vorschriften für Eisenbetonbauten," *SBZ* 99, no. 5:55-59.

1932   "Zum Entwurf der neuen schweizerischen Vorschriften für Eisenbeton" (Korrespondenz), *SBZ* 99, no. 10:125-126.

1932   "Ueber Erdbebenwirkung auf Hochbauten," *SBZ* 100, no. 24:309-311.

1932   "Die Wandlung in der Baukonstruktion seit 1883," *SBZ* 100, no. 27:360-364.

1932   "Théories des dalles à champignons," *Premier congrès international des ponts et charpentes de Paris*, Final Report, Paris, pp. 197-208.

1932   "Die Verhältniszahl n = 15 und die Zulässigen Biegungsspannungen," (discussion), *Beton und Eisen* 31, no. 1:10.

1933   (With M. Turretini) "Neues Bankgebäude der Schweizer. Kreditanstalt an der Place Bel Air in Genf," *SBZ* 101, no. 4:47.

1933   "Die Zürcher Sport-und Grünanlagen im neuen 'Sihlhölzli': Konstruktives," *SBZ* 101, no. 9:103-104.

1934   Ponts-voûtes en béton armé, *Bulletin du ciment* 2, no. 8:2-6.

1934   "Gekrümmte Eisenbeton-Bogenbrücken," *SBZ* 103, no. 11:132-133.

1935   "Flachdächer ohne Gefälle," *SBZ* 105, no. 15:175-176.

1935   "Die schweizerischen Normen für Eisenbeton von 1935," *Der Bauingenieur* 16, no. 47/48:481-485.

1935   "Verstärkung einer Eisenbetonkonstruktion," *SBZ* 105, no. 11:130-132.

1935   "La construction des ponts en béton armé, envisagée au point de vue esthétique," *Le Génie civil* 106, no. 11:262.

1935   "Ponts-voûtes en béton armé: de leur développement et de quelques constructions spéciales exécutées en Suisse," *Travaux: Architecture, Construction, Travaux-Publics* 19, no. 26:64-71.

1935   "The Construction and Aesthetic of Bridges," *The Concrete Way* 7, no. 6 (May/June):303-309 (translation with some modifications of article in *Le Génie civil* 106, no. 11:262).

1936   "Einige neuere Eisenbetonbrücken," *SBZ* 107, no. 15:157-162.

1936   "Der Ausbau des Quai Perdonnet in Vevey," *SBZ* 108, no. 15:159-161.

1937   "The Modular Ratio," *Concrete and Constructional Engineering*, Sept. 1937, pp. 517-520.

1938   "Aktuelle Fragen des Eisenbetonbaues," *SBZ* 111, no. 1:1-5.

1938   "Ueber Eisenbeton-Brücken mit Rippenbögen unter Mitwirkung des Aufbaues," *SBZ* 112, no. 24:287-293.

1939   "Evolution de la construction des ponts en béton armé," *Bulletin technique de la suisse romande* 65, no. 7:85-91 (partial reprint of 1935 article in *Le Génie civil* with some remarks by S. Giedion)

1943   "Kongressaal," *SBZ* 121, no. 23:282-283.

SWITZERLAND

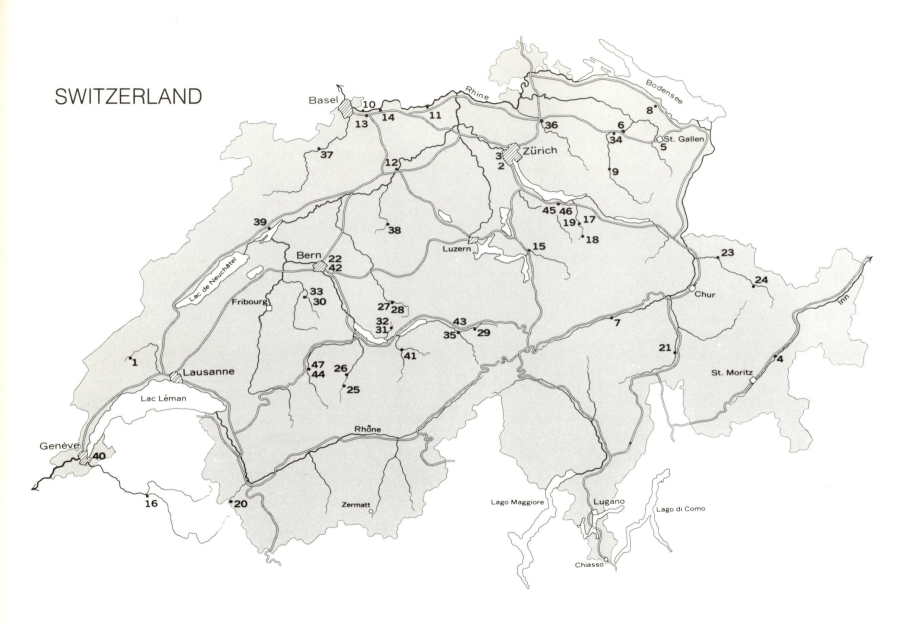

Basel

Rhine

Bodensee

10

14   11

13   36   8

37   3   Zürich   6   St. Gallen
12   2   34
5

9

45 46
39   19   17
38   18

Bern 22   Luzern   15   23
42
33   24
Fribourg 30   27 28   43   Chur
Lac de Neuchâtel
32   35   29   7
31

1   41   21
47   26
Lausanne 44   St. Moritz
25   4

Lac Léman

Genève 40   Rhône   Inn

16 20   Zermatt   Lago Maggiore Lugano Lago di Como

Chiasso

Map by R. L. Williams

Appendix B The Major Bridges of Robert Maillart

Number	Date	Place (Name)	Crossing	Type[a]
1	1896	Pampigny	Le Veyron Brook	a
2	1899	Zurich (Stauffacher)	Sihl River	b
3	1901	Zurich	Hadlaub Street	c
4	1901	Zuoz	Inn River	d
5	1903	St. Gallen	Steinach Brook	e
6	1904	Billwil	Thur River	d
7	1905	Tavanasa[b]	Rhine River	d
8	1907	Aach	Railroad	c
9	1909	Wattwil	Thur River	d
10	1910	Wyhlen	Unterwasser Canal	f
11	1911	Laufenburg	Rhine River	e
12	1912	Aarburg	Aare River	a
13	1912	Augst-Wyhlen	Rhine River	c
14	1912	Rheinfelden	Rhine River	e
15	1913	Ibach	Muota River	g
16	1920	Marignier	Arve River	a
17	1923	Wäggital[c]	Flienglibach	h
18	1924	Wäggital	Ziggenbach	a
19	1924	Wäggital	Schrähbach	h
20	1924	Châtelard	Eau-Noire	f
21	1925	Donath	Valtschielbach	h
22	1930	Bern (Lorraine)	Aare River	e
23	1930	Schiers	Salginatobel	d
24	1930	Klosters	Landquart River	h
25	1931	Adelboden (Spital)	Engstligen River	h
26	1931	Frutigen (Ladholz)	Engstligen River	h
27	1931	Schangnau	Hombach	h
28	1931	Schangnau	Luterstalden Brook	h
29	1932	Nessental	Triftwasser Brook	i
30	1932	Schwarzenburg	Rossgraben	d
31	1932	Habkern	Traubach	h
32	1932	Habkern	Bohlbach	h
33	1933	Hinterfultigen	Schwandbach	h
34	1933	Felsegg	Thur River	d
35	1934	Innertkirchen	Aare River	d
36	1934	Wülflingen-Winterthur	Töss River	h

Number	Date	Place (Name)	Crossing	Type[a]
37	1935	Liesberg	Birs River	c
38	1935	Huttwil	Railroad	c
39	1936	Twann	Twannbach	d
40	1936	Vessy	Arve River	d
41	1937	Gründlischwand	Lütschine River	c
42	1938	Bern	Weissenstein Street	c
43	1938	Wiler	Gadmerwasser	d
44	1940	Laubegg	Simme River	i
45	1940	Altendorf	Railroad	c
46	1940	Lachen	Railroad	d
47	1940	Garstatt	Simme River	d

[a] All bridges of reinforced concrete except as noted.
[b] Bridge destroyed by avalanche.
[c] Bridge later replaced.

Key to Bridge Type

a. Hingeless arch
b. Unreinforced, three-hinged arch
c. Continuous beam
d. Three-hinged, hollow-box arch
e. Concrete-block arch
f. Continuous hollow-box beam
g. Hollow-box, cantilever beam
h. Deck-stiffened arch
i. Beam

Notes

Robert Maillart's private papers are (1978) in the hands of Marie-Claire Blumer-Maillart of Zurich. His business documents are in various cantonal and city archives in Switzerland. Whenever possible, copies have been placed in the Maillart collection at Princeton University Library, indicated in the notes by the abbreviation PMA. As of 1978, this material comprises copies of about ten thousand such documents, including private letters and other papers, business correspondence, plans, calculations, and photographs.

All translations are by the author unless otherwise stated.

The major Swiss civil engineering journal, *Schweizerische Bauzeitung* is abbreviated *SBZ*.

Chapter 1

1 "Généalogie de la Famille Maillart," a partially completed family document signed by Ed. Senemaud, April 1, 1868, 9 pages, PMA. There is nothing in this document about any of the Belgian descendants of the nineteenth century. On Sept. 19, 1908, J. Kaufmann, an architect of Zurich, won second prize in the competition for a schoolhouse in Monthey with a project entitled "Colin-Maillard," *SBZ* 52, no. 12 (Sept. 19, 1908): 159.

2 *Biographie Nationale de Belgique*, 1895, Vol. 13: 171-186.

3 "Maillart Genealogy," private manuscript, Bern, belonging to a branch of the Maillart family. See also *Dictionnaire Historique et Biographique de la Suisse*, 1928, Vol. 4: 638.

4 The Registry of Baptisms in the District of Bern, Vol. 4, May 21, 1872 (the date Maillart was baptized) shows his father Edmond was still a Belgian citizen at the time.

5 Rosa (1865-1950); Paul (1866-1936), whose daughter Ella became a well-known writer in both English and French; Marguerite (1867-1873); Alfred (1869-1941); Robert (1872-1940); and Maximilian Charles (1873-1942).

6 Maillart's school records copies in PMA.

7 Reyner Banham, *Theory and Design in the First Machine Age*, London, Architectural Press, 1960, p. 14.

8 Ibid., p. 141. Banham describes the influence of Semper on H. P. Berlage, the Dutch architect who had studied in Zurich in the 1870's.

9 "Baukonstructionslehre," Maillart's class notes from the course by Professor Recordon, PMA.

10 Wilhelm Oechsli, *Geschichte der Gründung des Eidg. Polytechnikums mit einer Übersicht seiner Entwicklung, 1855-1905*. Frauenfeld, Huber & Co., 1905, pp. 177-178.

11 Carl Culmann, *Die Graphische Statik*, 1st ed., Zurich, von Meyer and Zellers Verlag, 1869. The diagram in Fig. 1-1a appears in the 2nd ed., 1875, p. 594.

12 Karl Wilhelm Ritter was from Altstätten (Canton St. Gallen). He graduated from the ETH in 1868, where he became an assistant. In 1870 he became a lecturer in engineering science, and in 1873 was named professor at the Technical University in Riga. In 1882 he was called back to Zurich as a professor, following the death of Carl Culmann in December 1881. From 1887 to 1891 Ritter was director of the ETH. A long illness forced his early retirement in 1905; see Oechsli, *Geschichte der Gründung*, pp. 317, 335-352.

13 Wilhelm Ritter, *Der Brückenbau in den Vereinigten Staaten Amerikas*, Zurich, Verlag von Albert Raustein, 1895. This book was occasioned by Ritter's visit to the Chicago World's Fair in 1893, and some of its contents can be seen in Maillart's "Brückenbau" class notes from Ritter's course, 1893, PMA.

14 R. Maillart, "Brückenbau," May 17-26, 1893. The system number four, arches stiffened by truss work (*der durch Fachwerke versteifte Bogen*), contrasts in its clear combined form with the system number one which looks thoroughly empirical and unstiffened. Ritter gave the Paris Ivry bridge over the Seine as an example of system one and the Davenport bridge (75-meter span) over the Mississippi as an example of system four; it is this bridge Maillart called marvelous.

15 Maillart, "4021/88 Statische Berechnung, Eisenbetonbrücke über den Flienglibach," 1923, p. 5, PMA.

16 Fritz Stüssi, *Othmar H. Ammann*, Basel, Birkhauser Verlag, 1974. Clearly Maillart and Ammann were outstanding, and whether or not they were the most outstanding is a matter of personal judgment.

Chapter 2

1 For the present study, the only reinforced concrete work of significance in Switzerland before 1894 was the Monier arch bridge at Wildegg near Zurich, built in 1890 by the German, Wayss. The 3.9-meter-wide arch had a span of 37.22 meters, a crown rise of 3.5 meters, and thicknesses of 20 cm at the crown and 65 cm at the springing lines. See *SBZ* 17, no. 11 (March 14, 1891): 66-68.

2 Peter Collins, *Concrete: The Vision of a New Architecture*, London, Faber, 1959, pp. 60-61.

3 Ibid., p. 65-72. Collins gives an outline of the origins of concrete construction and describes the early developments by Ransome in America, Monier and the Coignets in France, and Wayss (through the Monier patents) in Germany. Hennebique was by no means the only important figure and he certainly drew on the work of others.

4 Paul Christophe, *Le Béton armé et ses applications*, 2nd ed., Paris, Librairie Polytechnique, 1902.

5 "Concrete and Concrete-Steel," *Transactions of the American Society of Civil Engineers*, Vol. 54, Part E, New York, 1904.

6 Edwin Thatcher, "Concrete and Concrete-Steel in the United States," ibid., p. 437.

7 Fritz von Emperger, "Concrete-Steel Bridges," ibid., p. 522.

8 Ibid., p. 526.

9 A. Favre, *SBZ* 25, no. 5 (Feb. 2, 1895):31-32; S. Rappaport, *SBZ* 29, no. 9 (Feb. 27, 1897): 61-63.

10 Wilhelm Ritter, "Die Bauweise Hennebique," *SBZ* 33, no. 5 (Feb. 4, 1899): 41-43; no. 6 (Feb. 11, 1899): 49-52; no. 7 (Feb. 18, 1899): 59-61.

11 S. de Mollins, "Le béton armé Système Hennebique" (Korrespondenz), *SBZ* 33, no. 12 (March 25, 1899): 109.

12 Robert Maillart, "Das Hennebique-System und seine Anwendungen," *SBZ* 37, no. 21 (May 25, 1901): 225-226.

Chapter 3

1 For the correspondence between Maillart and Pümpin and Herzog, see documents from 1894-1896, PMA.

2 For the official changes in Maillart's addresses, see *"Dienstbüchlein* für Robert Maillart, 1891-1919," PMA.

3 For a discription of the opening, see "Inauguration du Bière-Apples-Morges," *L'Ami de Morges*, July 3, 1895, PMA.

4 For details of the bridge near Pampigny, see "Pont de 6.00 m. sur le Veyron," report dated December 28, 1894, PMA.

5 The cost of the Pampigny bridge, 100 Swiss francs per square meter of roadway, is found in the "Estimation: Pont Route de 6.00 m. sur le Veyron," PMA. It was a reasonable cost for a small bridge at that time (see Table 4-2). There is no date but presumably it was late 1894. The drawings with this estimate show a railway bridge, but the estimate has a heading, "Route Cantonale No. 58." It is not clear, therefore, whether the Maillart design was for both the roadway and the railway bridges. I assume that it was for the railway only.

6 "Inauguration du Chemin de Fer Apples-L'Isle," *L'Ami de Morges*, Sept. 16, 1896, PMA.

7 Pümpin und Herzog, "Zeugnis" (for Robert Maillart), Bern, Nov. 7, 1896, PMA.

8 "Auszug aus dem Protokoll des Vorstandes," Zurich, Dec. 15, 1896, PMA.

9 V. Wenner, "Presentation to the Zurich Branch of the SIA [Society of Swiss Engineers and Architects] on Feb. 8, 1899," *SBZ* 33, no. 9 (March 4, 1899): 82.

10 See Chapter 1, note 12.

11 W. Ritter, "Consultation on the Design of a Stauffacher Bridge in Zürich," Aug. 9, 1898, p. 1, PMA.

12 Ibid., p. 6.

13 Ibid., p. 9.

14 V. Wenner, "Memorandum to the Building Commission," Aug. 13, 1898, PMA.

15 V. Wenner, "Presentation to the Zurich Branch of the SIA."

16 W. Ritter, "On the New Construction of the Coulouvrenière Bridge in Geneva," *SBZ* 27, no. 14 (April 11, 1896): 100-101.

17 Indeed, the arch was built essentially as blocks. The builder cast the arch in a series of narrow strips running the width of the bridge. In this way, the wooden scaffold could settle under the loading of the first strips, which could move downward freely since initially only every other strip was cast.

18 Even though Maillart's design appears to have been significantly less costly, the entire project, including roadways, was criticized as unnecessary and expensive and was nicknamed the "Spekulanten-Brücke"; see *Tages-Anzeiger für Stadt und Kanton Zürich*, June 3, 1899.

19 V. Wenner, "Presentation to the Zurich Branch of the SIA."

20 M. Roš, "Robert Maillart:Ingenieur, 1872-1940," *Schweizerischer Verband für die Materialprüfungen der Technik*, Zurich, April 1940. On this document, see below, Chapter 4, note 2.

21 "Auszug aus dem Protokolle des Stadtrates von Zürich," Vol. 30, August 1899, PMA.

22 Hans Studer, "Steinerne Brücken der Rhätischen Bahnen," *Schweizerische Ingenieurbauten in Theorie und Praxis, International Kongress für Brückenbau und Hochbau*, Zurich, 1926. Studer states that the calculations were new, that they followed Ritter's method, and that they were made in 1900. He does not mention Maillart. However, Maillart sent picture postcards of the Solis bridge twice to his eldest son, Edmond, and once to his wife; on two of these he noted that he had made the calculations for this bridge. See Maillart's card (Fig. 3-4) of Saturday, Sept. 7, 1903, to Maria on which he noted: "Ecco un bel ponte! J'y suis passé pour venir à l'Engadine. J'en ai fait les calculs. Rob." From the archives of the Rhätische Bahn in Chur it appears that Studer first made the calculations which were presented in a drawing dated Feb. 1901. These were gone over by an engineer named Mantel who wrote a report dated March 16, 1901. Then Froté and Westermann apparently were asked to make a new set of calculations which they submitted with a report dated May 4, 1901 probably written by Maillart. There then followed a revised design, probably by Studer, presented in a drawing dated June 1901. Copies of these reports and drawings are in PMA.

23 "Innbrücke Zuoz," drawing and specifications signed by E. Ganzoni, July 1900. The cost estimate appears on a document signed by Carl Peterelli, canton engineer, Nov. 1900. Both PMA.

24 W. Catrina, *Die Entstehung der Rhätischen Bahn*, Zurich, 1972, p. 124.

25 Froté and Westermann [Maillart], "Report for a bridge over the Inn near Zuoz designed as an arch bridge of reinforced concrete," Zurich, Aug. 13, 1900, PMA.

26 Ibid., p. 2.

27 Ibid., p. 3.

28 A. Schucan, to Froté and Westermann, Sept. 18, 1900, PMA. The reason for lowering the price was that the wing walls were now to be built differently. Also, the letter stipulated that the abutments were to be completed before the winter.

29 For a description of Bonna's works, see Paul Christophe, *Le Béton armé et ses applications*, Paris, 1902, pp. 6, 22, and especially pp. 274-275, where Christophe shows a photo of Bonna's arch bridge over the Gers River at Auch. This bridge had been described in the French journal *Le Ciment* in 1899, p. 161. Bonna's bridge is one of the few shown by Christophe in which the structure is fully exposed; it has a modern look and Maillart undoubtedly knew about it. In his historical review of reinforced concrete, pp. 5-6, Christophe mentions Bonna as one of the leaders in France.

30 Christophe, *Le Béton armé*, p. 6.

31 Froté and Westermann in Zurich to Maillart in Paris, Sept. 25, 1900, PMA.

32 E. Ganzoni in Samedan to W. Ritter in Zurich, Feb. 3, 1901, PMA. Possibly Ganzoni sent the design with this letter; but more likely it had been sent earlier and the letter was a reminder.

33 W. Ritter in Zurich to E. Ganzoni in Samedan, March 5, 1901, PMA.

34 "Bauvertrag, Zuoz," July 10, 1901. Ritter's report is not with these documents, but it is referred to in this building contract, PMA.

35 W. Ritter, "Report on the Load Test of the New Inn Bridge near Zuoz," Zurich, winter 1901-1902, pp. 11-12, PMA.

36 A 4-meter width of the solid arch has a weight of 9.5 tons for every meter of length, and the weight of the roadway deck is about 2.5 tons per meter (including some allowance for the vertical cross walls that connect deck and arch). Thus, a 4-meter-wide section of the Stauffacher carries a total weight of 12 tons per meter on an arch section of 3.8 m² (0.95 m depth times 4 m width). By contrast, the Zuoz bridge, which is about 4 meters wide, has a total weight of 5.1 tons per meter to be carried by an arch section of 2.05 m². Calculations for the stresses give 15.8 kg/cm²

for Stauffacher and 12.4 kg/cm² for Zuoz. Stauffacher is higher because the load is 2.35 times higher even though its arch is 1.85 times bigger. Thus, the stresses are directly proportional to the load and inversely proportional to the arch area, i.e., 15.8/12.4 and 2.35/1.85 both equal 1.27. This present argument does not include the influence of live loads (cars, people, snow, etc.) which would only show the Zuoz to be even more efficient.

These numbers were derived as follows:

(a) A parabolic arch with a load uniformly distributed along a horizontal projection of its span pushes against its abutments with a horizontal force $H = w\ell^2/8f$ where w is the load in tons per meter, ℓ is the span in meters, and f the rise of the arch at its crown, also in meters. Then for Stauffacher, $H_s = 12 \times 40^2/8 \times 4 = 600$ tons; and for Zuoz, $H_z = 5.1 \times 40^2/8 \times 4 = 255$ tons.

(b) For flat arches the axial force at the quarterspan $N \simeq H$ and the resulting stress will thus be approximately H/A, so that $\sigma_s = 600/3.8 = 158$ T/m² = 15.8 kg/cm² and $\sigma_z = 255/2.05 = 124$ T/m² = 12.4 kg/cm². These calculations approximate both arches as 40 meters in span and 4 meters in rise; using the actual dimensions leads to slightly different numbers, but does not alter the argument in any way.

37 This procedure leads to loadings different from those described in note 36. If Maillart had really achieved his loading sequence, then the bottom arch would have had more stress than we have computed. As Maillart's official calculations for the Salginatobel and later works show, however, he followed the loading assumptions made in note 36 and thus probably used the constructional argument only as a basis for reducing scaffolding. See, for example, D. P. Billington, "Explanation of the Arch Analysis," in J. F. Abel, ed., *Background Papers, Second National Conference on Civil Engineering: History, Heritage, and the Humanities*, Princeton University, October 4-6, 1972, published by the Department of Civil Engineering, Princeton University, 1972, pp. 91-103.

38 The stone covering does serve to protect the concrete from weathering even though good concrete normally does not need such protection.

39 R. Favre, "Die Erneuerung von zwei Maillart-Brücken," *SBZ* 87, no. 17 (April 24, 1969): 313-319.

40 Emil Mörsch, *Concrete-Steel Construction* (translated by E. P. Goodrich from the 3rd [1908] German ed.), New York, Engineering News Pub. Co., 1910, pp. 271-280. This book was originally published in German in 1902, 2nd ed. 1906.

41 Ibid., p. 279. See also E. Mörsch, "Die Isarbrücke bei Grünwald," *SBZ* 44, no. 23 (Dec. 3, 1904): 263-267; and 44, no. 24 (Dec. 10, 1904): 279-283.

42 Quoted in R. Banham, *Theory and Design in the First Machine Age*, London, Architectural Press, 1960, p. 94.

43 Ibid., p. 97.

44 Maillart, "Notes for a Slide Lecture in Basel," 1938, PMA.

45 M.-C. Blumer-Maillart, "Recollections of My Father," 1978, PMA.

46 *Zürcher Wochen-Chronik*, Vol. 3, no. 36, p. 291, PMA.

Chapter 4

1 This treatment gives the bridge an impressive and unique appearance somewhere between stone and concrete. In fact, it was singled out for special praise by two Americans in 1916, who made a trip to Europe to study bridges, and devoted more space to the Steinach than to any other European bridge in the section on surface treatment of concrete. See Philip Aylett and P. J. Markmann, "European Concrete Bridges," *Reinforced Concrete Construction*, Vol. 3: *Bridges and Culverts*, ed. G. A. Hool and F. C. Thiessen, New York, McGraw-Hill, 1916, pp. 629-630. In fact, they gave special attention to twenty-one bridges (i.e., showed more than one photograph) of which fourteen were Swiss, including four by Maillart (Stauffacher, Steinach, Tavanasa, and Aarburg). Unfortunately, they did not mention designers, so Maillart's name never appears.

2 M. Roš, "Robert Maillart: Ingenieur, 1872-1940," *Schweizerischer Verband für die Materialprüfungen der Technik* (Swiss Association for Materials Testing), Zurich, April 1940. This special publication commemorating Maillart just after his death contains a list of his works. Those completed between 1902 and 1920 were carried out by his design-construction company. The Stauffacher and Zuoz bridges are not included (even though Roš shows a photo of the former and does mention both in his brief article on Maillart). Maillart built many minor works not listed by Roš.

3 The comparisons are as follows: spans, Zuoz 38.25 m, Billwil 35 m; rise between springing lines and crown, Zuoz 3.6 m, Billwil 4.2 m; roadway width, Zuoz about 3.6 m, Billwil 3.56 m. The Zuoz dimensions are from R. Favre, "Die Erneuerung von zwei Maillart-Brücken"; and for Billwil from Bersinger (cantonal engineer for St. Gallen), "Die neue Strassenbrücke über die Thur bei Billwil-Oberbüren, Kanton St. Gallen," *SBZ* 44, no. 14 (Oct. 1, 1904): 157-159.

4 Based on the live loads of 300 kg/m² and a truck of 12 tons, and an assumed uniform dead load approximated by the weight of the quarterspan section, we can compute the axial forces at the crown and at the springing line as follows:

(a) Cross sections:
Crown:

Deck slab	4.04 × 0.12	= .49
Walls	0.48 × (0.8 − 0.28) =	.25
Arch slab	3.20 × 0.16	= .51
Total area A_c		= 1.25 m²

Quarterspan:

Deck slab		= .49
Walls	0.48 × (1.8 − 0.28) =	.73
Arch slab		= .51
Total area A_q		= 1.73 m²

Springing:

Arch slab	3.20 × 0.6	= 1.92 m²
Total area A_s		= 1.92 m²

(b) Horizontal thrust
Dead load concrete = 1.73 m² × 2.5 T/m³ = 4.33 T/m
Dead load surface = 3.56 × .16 × 2.5 T/m³ = 1.42 T/m

Live load uniform = 3.56 × .3 T/m² = 1.07 T/m
Live load truck = 12-ton concentrated load

$$H_u = \frac{w\ell^2}{8f} = \frac{(4.33 + 1.42 + 1.07)35^2}{8(4.2)} = 250\ T$$

$$H_c = \frac{P\ell}{4f} = \frac{(12)35}{4(4.2)} = 25\ T$$

$$H_T = 250 + 25 = 275\ T$$

(c) Vertical shear and axial thrust at springing

$$\varphi = \arctan \frac{4f}{\ell} = \arctan(.48) = 25.6°$$

$$V_T = (6.83)35/2 + 12/2 = 126\ T$$
$$N_T = V \sin\varphi + H_T \cos\varphi = 126(.432) + 275(.902) = 303\ T$$

(d) Stresses
Crown: $\sigma = H_T/A_c = 275/1.25 = 22$ kg/cm²
Springing: $\sigma = N_T/A_s = 303/1.92 = 15.8$ kg/cm²

These values compare to 23.5 and 15.1 given in the article by Bersinger. Considering our approximations, this correspondence is good and confirms our explanations of Maillart's method. These are low stresses, emphasizing the conservative nature of the design. Bersinger's results come from Maillart; see note 15.

5 Maillart, *Bogenträger aus armiertem Beton*, Patent No. 25712, Swiss Patent Office, Bern, Feb. 18, 1902.

6 *Geographisches Lexikon der Schweiz*, 1902, Vol. 1: 263. Zuoz by contrast had 105 houses and 425 souls in the year 1910; see ibid., Vol. 6: 851. (Because the *Lexikon* was completed over an 8-year period, Vol. 1 represents the situation up to 1902 and Vol. 6 up to 1910.)

7 Bersinger, "Die neue Strassenbrücke über die Thur bei Billwil-Oberbüren, Kanton St. Gallen."

8 "Cost Estimate for a Steel Bridge over the Thur," St. Gallen, Jan. 1898, PMA.

9 Maillart & Cie in Zurich to canton engineer Bersinger in St. Gallen, May 30, 1903, PMA.

10 Ibid.

11 W. Ritter in Zurich to canton engineer Bersinger in St. Gallen, June 17, 1903, PMA.

12 Bersinger, "Die neue Strassenbrücke über die Thur bei Billwil-Oberbüren, Kanton St. Gallen."

13 Maillart & Cie in Zurich to canton engineer Bersinger in St. Gallen, Aug. 7, 1903, PMA.

14 "Wettbewerb für die neue Utobrücke," *SBZ* 44, no. 7 (Aug. 13, 1904): 76-80.

15 Maillart & Cie, *"Thurbrücke Billwil, Statische Berechnung,"* Aug. 6, 1903, included with the letter of Aug. 7 (see note 13), PMA.

16 Maillart & Cie to Bersinger, Aug. 7, 1903, p. 2.

17 W. Ritter, "Report on the Load Test of the New Inn Bridge near Zuoz," winter 1901-1902, p. 9, PMA.

18 W. Ritter, "Consulting Report on the Design of Mr. Maillart & Cie for a Thur Bridge by Billwil," Zurich, Aug. 22, 1903, PMA.

19 Maillart & Cie Zurich to canton engineer Bersinger in St. Gallen, Aug. 28, 1903, PMA.

20 "Die Kollaudation der Thurbrücke bei Billwil (Oberbüren)," *Wyler Bote*, Wil, April 16, 1904.

21 M.-C. Blumer-Maillart, "Recollections of My Father," 1978, PMA.

22 Maillart to Maria, Zurich, Sept. 9, 1903, PMA.

23 Maillart to Schucan, Sept. 26, 1903, PMA.

24 Maria Maillart in Zurich to Bertha Maillart in Bern, Oct. 2, 1903, PMA.

25 "Wettbewerb für die neue Utobrücke," *SBZ* 44, no. 7 (Aug. 13, 1904): 76-80.

26 Maillart to Maria, Zurich, June 28, 1904, PMA.

27 Maillart to Maria, Zurich, Sept. 26, 1904, PMA.

28 Maillart, card to Maria, Tavanasa, Aug. 2, 1904, PMA.

29 The Zuoz span was estimated by Max Bill to be 30 meters; see M. Bill, *Robert Maillart*, 3rd ed., Zurich, Verlag für Architektur, 1969, p. 32. But according to Favre, who rebuilt the bridge, its span is 38.25 meters; see R. Favre, "Die Erneuerung von zwei Maillart-Brücken," *SBZ* 87, no. 17. It seems important to observe that at Billwil Maillart did not take the more dramatic, or more advanced choice of one 70-meter span, probably because of the low rise. Mörsch's 70-meter-span Isar bridge had a rise of 12.8 meters or a ratio of 5.5, whereas even with 35 meters, Billwil has a span-rise ratio of 8.4. A higher ratio means higher horizontal foundation loads and also higher forces in the arch.

30 S. Solca, "Die Rheinbrücken bei Tavanasa und Waltensburg," *SBZ* 63, no. 24 (June 13, 1914): 343-344. Bersinger, "Die neue Strassenbrücke über die Thur bei Billwil-Oberbüren," *SBZ* 44, no. 14 (Oct. 1, 1904) 157-159.

31 E. Mörsch, *Concrete-Steel Construction* (translated by E. P. Goodrich from the 3rd [1908] German ed.), p. 283. Mörsch was familiar with Maillart's work, not only because he was a professor at the ETH in Zurich, but also because he had been called in as a consultant by the Canton of Graubünden to check Maillart's Tavanasa design; see *SBZ* 63, no. 24 (June 13, 1914): 344.

32 More accurately, if the load P is the same and the span changes from ℓ_1 to ℓ_2, then the bending moment changes from $P\ell_1/4$ to $P\ell_2/4$. These moments give stresses computed from $\sigma = 6M/bd^2$, where M is the moment, b the section width, and d its depth. Thus, if σ and b are constant, then the depth must increase when ℓ increases as follows:

$$\sigma_1 = 6M_1/bd_1^2 = \frac{3}{2}\frac{P\ell_1}{bd_1^2}$$

$$\sigma_2 = \frac{3}{2}\frac{P\ell_2}{bd_2^2}$$

Since $\sigma_1 = \sigma_2$,

then $\dfrac{\ell_2}{\ell_1} = \dfrac{d_2^2}{d_1^2}$.

Thus, comparing Billwil and Tavanasa dimensions shows that $1.96/1.26 = 1.55 \simeq (.12/.1)^2 = 1.44$. This simplified analysis neglects the facts that the deck at Billwil is continuous, that the wheel loads are not single but double loads (2 wheels), and that different loads were specified for the two bridges. But the principle remains the same and explains Maillart's rationale for eliminating the third wall.

33 This simplified argument is based on the following approximate analysis:

(a) Consider a flat parabolic arch under its own dead weight only. Then the arch force at the crown is purely horizontal and very nearly equal to $H = w\ell^2/8f$, where w is the dead weight and equals the material density γ (pounds per cubic foot) times the cross sectional area A (square feet), ℓ is the span, and f is the rise at the crown.

(b) the stress in the crown cross section $\sigma = H/A = w\,\ell^2/8fA = \gamma A\ell^2/8fA = \gamma\ell^2/8f$, which for a span of 51.25 m and a rise of 5.7 m gives 13.8 kg/cm² (196 psi).

(c) Therefore, the stress depends only upon the material density, the span length, and the arch rise. Hence, dead load gives no criterion at all for dimensioning the arch. The stress of 196 psi is low.

34 The crown force at Tavanasa can be approximately checked by computing the section area at quarterspan:

Deck slab 360 × 13 = 4680
Walls   32 (220-30) = 6080
Arch slab 280 × 18 = 5040
                A_Q = 15,800 cm²
$w = \gamma A_Q = 2.5 \times 1.58 = 3.95$ T/m

$H = \dfrac{3.95(51.25)^2}{8(5.7)}$ = 229 T

This value is based on the assumption that the weight of the bridge is equal to the weight of the quarterspan section times the span length. The resulting force is only 3.7% low indicating a close approximation. The area of the section just next to the solid hinge section at the crown is approximated as:

Deck slab	=	4680
Walls 32 (82-31)	=	1630
Arch slab	=	5040
A_c	=	11,350 cm²,

so that the crown stress σ_c = 237.8/1.135 = 209 T/m² = 20.9 kg/cm².

The Billwil value is obtained by taking only the arch dead weight of 4.33 T/m from note 4 and computing H = 250 (4.33/6.83) = 159 T and σ_c = 22 (159/275) = 12.7 kg/cm². The quantity that determines stress in a given material is thus ℓ^2/f or 461 m for Tavanasa and 292 m for Billwil. The ratio is thus 461/292 = 1.58, whereas the computed ratio of stresses is 20.9/12.7 = 1.64 ≈ 1.58. Considering all the approximations made in arriving at the ratios, we can conclude that the check is very good.

35 At Billwil the cross-sectioned area A increases from 1.25 m² at the crown, to 1.73 m² at the quarterspan to 1.92 m² at the springing, while the axial force N from full load increases from 275 tons at the crown to 303 tons at the springing. The results of N/A for dead load alone show the stresses decreasing from crown to springing because live-load stresses make larger sections desirable away from the crown.

36 Even in large masonry works live loads are important, but because of the great weight of masonry and its restricted span lengths and building heights, live loads rarely play any noticeable role in the design of forms. One important exception occurred in the building of or modifications to the high Gothic cathedrals, where wind loads apparently caused distress which led to some changes of form. See R. Mark and R. J. Jonash, "Wind Loading on Gothic Structure," *Journal of the Society of Architectural Historians* 29 (Oct. 1970): 220-230.

37 This simplified argument is correct only when the distribution of the load results in a simple-beam bending moment diagram of exactly the same form as the centroidal axis of the arch. For a parabolic arch, a constant load uniformly distributed along a horizontal line achieves this condition of no bending, as does a catenary arch for the curved distribution of the arch load itself. In general, for any load there is one arch axis that achieves this state. Unfortunately, the arch thickness variation for any axial form creates a loading that can not exactly be made to produce a moment diagram of the same shape as that axial form. However, for flat arches, the simplified arguments used here are essentially correct for the purpose of explaining structural ideas.

38 Because of the dead load, forces are larger at the springing than at the crown. It would be correct to make the springing thicker than the crown as Mörsch did. Why Maillart did the reverse at Stauffacher is not clear to me. For the Stauffacher dimensions see V. Wenner, "Presentation to the Zurich Branch of the SIA," *SBZ* 33, no. 9 (March 4, 1899): 83; and for Isar see Mörsch, *Concrete-Steel Construction*, p. 273.

39 Josef Melan, "The Theory of Arches and Reinforced Concrete Arches in Particular," *Handbuch für Eisenbetonbau*, ed. Fritz von Emperger, Berlin, 1908, Vol. 1: 425.

40 That form would be called the deck-stiffened arch, culminating in the Schwandbach and Töss Bridges. See D. P. Billington, "The Deck-Stiffened Arch Bridges of Robert Maillart," *Journal of the Structural Division of the American Society of Civil Engineers* 99 (July 1973): 1527-1539. It is possible to see the Tavanasa as a thin arch stiffened by the walls and deck. At the quarterspan Maillart's arch is merely 18 cm compared, for example, to Hennebique's arch at Châtellerault which at the crown was 54 cm increasing to the springing. Considering the live load of w = 300 kg/m² over one half the span as shown in Figure 4-7a, then the quarterspan bending moment M_q = $w\,\ell^2/64$ = 39.4 mT, and a stress in the d = 18 cm thick arch of σ = $6M/bd^2$ = 6(39.4)/2.8(0.18)² = 2600 T/m² = 260 kg/cm² = 3690 psi, obviously far above allowable limits even today, and in 1905 impossible when the allowable stress was of the order of 36 kg/cm². The thickness required would be d^2 = 260(2.8)/6(394) = 0.324, d = 57 cm, which is of the same order as Hennebique's bridge (see P. Christophe, *Le Béton armé*, p. 272).

41 *Beilage zum Kantons-Amtsblatt*, no. 18, 1904, p. 464, PMA.

42 Maillart to Maria, Zurich, Sept. 27, 1905. The dates of construction stages have been estimated from dated construction photos in PMA.

Chapter 5

1 M. Roš, "Robert Maillart: Ingenieur, 1872-1940," *Schweizerischer Verband für die Materialprüfungen der Technik*, Zurich, April 1940.

2 *Beton-und Monierbau Aktien-Gesellschaft*, brochure, Düsseldorf, May 1956, pp. 10-11, PMA. These thin hingeless reinforced concrete arches are illustrated in the brochure, which shows one of the two original such bridges and the only one in Switzerland at least up to 1897, a 39-meter-span roadway bridge at Wildegg near Zurich built in 1890.

3 P. Collins, *Concrete: The Vision of a New Architecture*, London, Faber, 1959, p. 72.

4 A French code commission was formed in 1900, producing official regulations by 1906 (P. Collins, *Concrete: The Vision of a New Architecture*, pp. 74-75); and a preliminary German recommendation appeared in 1904, followed by regulations published on May 24, 1907 (Mörsch, *Concrete-Steel Construction*, pp. 317-362). In September 1902 the central committee of the Swiss Society of Engineers and Architects began studies to produce a Swiss code, for which a commission was formed in October. This commission included five members from Zurich listed in the following order: Engineer R. Maillard [*sic*], Engineer V. Wenner, Architect O. Pfleghard, Engineer Löhle, and Engineer G. Meyer. See *SBZ* 41, no. 14 (April 1903): 159, where the proposed provisional code is also published.

5 R. Maillart, "Neuere Anwendungen des Eisenbetons" ("New Applications for Reinforced Concrete"), *Protokoll der ordentlichen Generalversammlung, am 16 und 17. September 1904*, Basel, Verein schweizerischer Zement- Kalk- und Gypsfabrikanten, Appendix III, pp. 16-23.

6 Maillart's speech was noted in *SBZ* 44, no. 15 (Oct. 8, 1904): 177-178. The collapse of the building for the restaurant and Hotel Bären had been mentioned in *SBZ* 38, no. 9 (Aug. 31, 1901), p. 96.

7 Prof. F. Schüle had reported at St. Louis on Maillart's early designs for tanks; see "Concrete and Concrete-Steel," *Transactions of the American Society of Civil Engineers*, Vol. 54, Part E, New York, 1905, p. 551.

8 R. Maillart, "Neuere Anwendungen des Eisenbetons," Appendix III, p. 17.

9 For the competition on Chauderon see *SBZ* 39, no. 13 (March 29, 1902): 143-144. There is no record of Maillart competing for this major work because bids were sought before he founded his own firm.

10 R. Maillart, "Neuere Anwendungen des Eisenbetons," Appendix III, p. 17.

11 P. Christophe, *Le Béton armé*. For tanks see pp. 372-380, for pipes pp. 330-334, for chimneys p. 381, for piles pp. 180 and 325-327. Maillart refers in his 1904 talk to high chimneys in America, to concrete masts with wooden cores in Grenoble, to concrete piles in Hamburg, and to concrete railroad ties under development in Spain, America, Italy and even Switzerland.

12 R. Maillart, "Neuere Anwendungen des Eisenbetons," Appendix III, p. 21.

13 The major source for Maillart's writings translated into English is Max Bill's book *Robert Maillart*, where none of these earlier works appear. Bill presented seven of Maillart's articles, of which six originally appeared after 1925; for the seventh, "Die Sicherheit der Eisenbetonbauten," published in 1909, Bill provided no English translation.

14 R. Maillart, "Belastungsprobe eines Eisenbetonkanals," ("Load Test of a Reinforced Concrete Conduit"), *SBZ* 50, no. 10 (Sept. 7, 1907): 125-128. As Maillart put it, "The complete encasement of the profile in soil, thereby bringing into play both active and passive earth pressure, makes invalid the assumption of a fully compacted bearing and hence makes the results of a calculation based upon the theory of elasticity completely false."

15 "Die Sicherheit der Eisenbetonbauten," *SBZ* 53, no. 9 (Feb. 27, 1909): 119-120. The report reads as follows: "The introduction of a new construction material with new properties is of great significance, because thereby new solutions are made possible to the designers' double problem of form and materials. Reinforced concrete, in spite of being composed of well-known materials, can be seen as a new material because its properties are not merely the sum of those constituent materials; rather it has new properties. A principal question about the introduction of any new material is whether it is safe enough. Here it is important to distinguish between resistance to physical and chemical influences and safety under external loads.

"Experience shows that reinforced concrete is highly fire- and frostproof, and that the rusting of the steel bars is no problem, even when moisture is trapped in the concrete. Also, reinforced concrete offers protection against earthquakes, through the ratio of small mass to great material strength (in contrast to the bending strength of masonry) and through good continuity between all the parts of a building.

"Dynamic effects have no damaging influence on reinforced concrete. As railroad ties demonstrate, one finds no change in the structure after years of service. An overwhelming further proof is afforded by reinforced concrete piles, which withstand blows from the heaviest rams. With heavy use, even where steel structural members are embedded in the concrete, there is still no danger of a loss in bond between the concrete and steel, although such construction should, in general, not be designated as reinforced concrete. Dynamic influences are taken care of purely through a percentage increase in the static load.

"Our understanding of safety in the sense of structural analysis is also unclear for older materials. Quite different results are obtained depending upon whether the largest design stresses are compared to the ultimate stresses from test pieces or whether the service loads are compared with the ultimate loads. Thus, one can speak of a ten- to twentyfold safety factor at first sight for a masonry arch, if the factor is based upon multiplying the applied load at the most dangerous level. It is, however, a great deception when masonry is credited with such great safety. Before the compression strength of the materials is even approximately reached, destruction will occur by exceeding the tension strength. If, therefore, an arch is capable of absorbing heavy tension stresses by steel reinforcement, then safety is greatly increased, even if under normal conditions the reinforcement does hardly any work. The results of the Austrian arch tests twenty years ago are proof. The one-sided failure load of the reinforced concrete arch was twice as large as that of the masonry [Bruchstein] and plain concrete arches, in spite of the fact that its crown thickness was only half.

"The basic mistake made by the theoretician was applying to reinforced concrete the coefficients and methods used to measure the safety factor of older materials. The validity of learning at first hand from the material itself and its behavior under working loads was recognized much later. The testing laboratory in Zürich has been outstanding in this research.

"When it was assumed that the failure of a reinforced concrete beam would occur when the calculated compression stress reached the cube breaking strength, research showed that the former was often exceeded by the latter by a factor of more than two. With quite low-strength concrete of about 75 kg/cm² compression strength, a 2½ times safety factor will still be available when the edge fiber compression stress is 60 kg/cm². With steel construction the safety factor is seldom larger than 2½ because the compression members in steel structures lose their carrying capacity at stresses of from 2,500 to 2,800 kg/cm². If, therefore, a minimum strength of 160 kg/cm² is required for concrete, and because of the naturally less constant quality of concrete a higher safety factor is needed, then one may allow a compression of 40 kg/cm² for axial loading and a compression of from 60 to 80 kg/cm² for eccentric loading in compression and bending. Also, it must be realized that these last are not real stresses, but rather numbers for comparison that, on the basis of a certain method of calculation, give a useful measure of the safety. The difference between actual and calculated stresses of steel at failure is less. Also, measuring the applied loadings for reinforced concrete requires other apparatus than for steel construction. For example, the German requirements for the calculation of continuous beams with unrestrained supports frequently lead to completely false results, as shown by deflection measurements from completed loading tests. Reinforced concrete is solidly built into its supports, and one must not ignore the bending moments that come into these supports. In any case, these relationships cannot be calculated exactly, and it is therefore better to pay attention now to the visible demonstrations gained in building experience.

"Beam theory is completely invalid for the analysis of slab structures. No previously known material can be used to build such structures on a large scale because stone has too low a tensile strength, and steel or wood can be used only in the direction of the rolling or the grain respectively. If, for analysis, one splits up such reinforced concrete slab structure into strips, the results are useless because the plate will be much thicker than necessary. Rather, it is essential to study the countless possible cases by measuring the deflections of appropriate test structures before one can hope to establish generally valid rules. Only some things will be singled out here from the many conclusions possible. These few principles have been well established: [then follow the five principles given here in the text].''

16 P. Christophe, *Le Béton armé*, pp. 572-573. The method preferred by Christophe was in general use before 1909; see F. E. Turneaure and E. R. Maurer, *Principles of Reinforced Concrete Construction*, 2nd ed., New York, Wiley, 1909, pp. 128-130, and E. Mörsch, *Concrete-Steel Construction*, pp. 59-60. Mörsch already recognized that something was not quite correct about the German code (*Leitsätze*) of 1904.

17 Turneaure and Maurer, *Principles of Reinforced Concrete Construction*, 5th ed., New York, 1932, p. 191. Two physical facts are crucial here: first, the theoretical design used by Christophe led to much heavier columns because it gave so little credit to the steel; and second, the theoretical analysis did not recognize that the column will shorten much more than its elastic deformation because of creep. The first fact led to waste of materials and the second to the cracking of structures where designers failed to consider potentially large deformations. Part of Christophe's objection to Hennebique's ideas lay in the fact that the latter had made assumptions for beam analysis that violated statics. But Hennebique had designed columns by calculating separately the capacities of the steel and of the concrete; it was this idea that formed one basis of the 1971 American code.

Chapter 6

1 O. Zehnder, "Die Aare-Brücke bei Aarburg," *SBZ* 62, no. 4 (July 26, 1913): 45-49.

2 F. Heitz, *Aarburg*, Bern, Verlag Paul Haupt, 1965. This little book gives a brief history of the town along with numerous photographs and reproductions of art works. The pictures emphasize its location at a bend in the Aare River and several clearly show Maillart's arch bridge.

3 Zehnder, "Die Aare-Brücke bei Aarburg": "eine Brücke mit einem Pfeilereinbau . . . befriedigte ästhetisch nicht. . . ."

4 This fact, to be made more explicit in Chapter 9 on deck-stiffened arches, arises from the relationship between absolute stress and relative stiffness as follows:

(a) Assume a fixed-base column connected at its top to two identical beams both fixed at their far ends (see sketch).

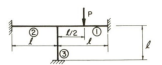

(b) Assume that each member has the same width b while members (1) and (2) have a depth h_1 and member (3) a depth $h_3 = \beta h_1$.

(c) Then the load P will produce a moment at the column top $M_3 = D_3(P\ell/8)$ where D_3, a distribution factor, is equal to the stiffness of (3) relative to the sum of the stiffness of (1) + (2) + (3) or

$$D_3 = \frac{4EI_3/\ell}{4EI_1/\ell + 4EI_2/\ell + 4EI_3/\ell} = \frac{\beta^3}{2 + \beta^3}$$

where $I_1 = bh_1^3/12 = I_2$, $I_3 = bh_3^3/12$, and E is constant throughout.

(d) The maximum stress at the column top will be

$$\sigma_3 = \frac{6M_3}{bh_3^2} = \frac{6}{bh_3^2} \left(\frac{\beta^3}{2 + \beta^3} \right) \frac{P\ell}{8}, \text{ or}$$

$$\sigma_3 = \frac{\beta}{2 + \beta^3} \sigma_f, \text{ with}$$

$$\sigma_f = \frac{6}{bh_1^2} \frac{P\ell}{8}$$

(e) If σ_f is held constant, the question then is to determine how the stress σ_3 varies when the relative stiffness of the column varies with respect to the beam, i.e., set

$$\frac{d\sigma_3}{d\beta} = 0, \quad \text{from which } \beta = 1.$$

Thus, as long as $\beta < 1$, a decrease in h_3 (hence β) leads to a decrease in σ_3, whereas when $\beta > 1$, a decrease in h_3 leads to an increase in σ_3.

(f) For the Aarburg Bridge the columns have $h_3 = 25$ cm and the deck ribs have $h_1 = 47$ cm (neglecting the extra stiffening influences of the deck slab which only makes the argument even stronger). Thus, $\beta = 0.53$ and $\sigma_3 = 0.246\sigma_f$; where $h_3 = 20$ cm then $\beta = 0.425$ and $\sigma_3 = 0.204\sigma_f$. But if we begin with $h_3 = 94$ cm, $\beta = 2$ and $\sigma_3 = 0.2\sigma_f$ (about the same as when $h_3 = 20$ cm) and then reduce h_3 to, say, 80 cm, then $\beta = 1.7$ and $\sigma_3 = 0.246\sigma_f$ or as much of an *increase* in σ_3 as the reduction in h_3 from 25 to 20 cm caused a *decrease*! The conclusion from this analysis would be that if the designer begins with heavy columns, then making them thinner would be less safe; whereas if the designer begins with thin columns as Maillart did, then making them thicker would be less safe (in the sense of increasing stresses). We shall meet this idea again in Chapter 9 with the deck-stiffened arches.

5 See, for example, the two-part article in the *Oltner Tagblatt*, Olten, April 5-6, 1968; and *The Technical Report*, Ingenieurbüro W. Schalcher, Zurich, Feb. 16, 1968, PMA.

6 R. Maillart, "Zur Berechnung der Deckenkonstruktionen," *SBZ* 59, no. 22 (June 1, 1912): 295-299.

7 R. Maillart, "Zur Berechnung der Deckenkonstruktionen," p. 296. For an abstract of Schüle's lecture, see *SBZ* 58, no. 21 (Nov. 11, 1911): 275, where it is noted that the discussion took place among Max Ritter, Maillart, and Schüle, though no details are given.

8 This distinction breaks down for short-span heavily loaded bridges and for long-span lightly loaded roof structures, the former usually becoming slab-like structures and the latter frequently arches and curved shells in which geometry takes precedence over materials. Maillart's Chiasso shed roof, described in Chapter 7, is a good example of the latter.

9 Giedion's first writing on Maillart appeared in *Cahiers d'Art* 5 (1930): 146-152 and was later incorporated into his *Space, Time and Architecture: The Growth of a New Tradition*, 5th ed., Cambridge, Mass., Harvard University Press, 1967, pp. 450-476. See esp. p. 475: "He, as the humble servant of architects constructed a very large number of buildings which do not reveal that he had anything to do with them. He never encountered an architect who fully knew how to integrate his genius. Where he was great, he was alone." Years later, the Italian engineer P. L. Nervi would show that flat slabs could be given a highly visible quality by curved ribbing. See P. L. Nervi, *Aesthetics and Technology in Building*, Cambridge Mass., Harvard University Press, 1965, pp. 80-81. Maillart's column capitals do give his flat slabs a fine visual quality and Giedion is partly correct in assuming that architects working with Maillart did not much use that quality in their designs.

10 R. Maillart, "Deckenkonstruktionen," p. 296.

11 Maillart recommended a factor of $2.4g + 3.6p$, rather than the $3(g + p)$ that was used in Switzerland at the time. Thus, he was implying that the uncertainly in live load was 50% greater than in dead load. The American Code (ACI-318-71) uses factors similar to this except that the live load is given a factor only about 21% greater than dead load.

12 Ernest L. Ransome and Alexis Saurbrey, *Reinforced Concrete Buildings: A Treatise on the History, Patents, Designs and Erection of the Principal Parts Entering into Modern Reinforced Concrete Building*, New York, McGraw-Hill, 1912, pp. 16, 153, and 161-170.

13 P. Christophe, *Le Béton armé*. See especially the index, where the major systems receive the following number of references: Hennebique 27, Monier 23, Matrai 13, Bonna 12, Coignet 10, Ransome 10, Moller 9, Boussiron 8, Bordenave 7, Cottancin 7, Melan 7, Golding 6, Habrich 5.

14 Eric Schiff, *Industrialization without National Patents: The Netherlands, 1869-1912; Switzerland, 1850-1907*, Princeton, Princeton University Press, 1971, pp. 85-95.

15 Ibid., pp. 96-106.

16 The Maillart patents were issued by the Swiss Patent Office in Bern. There are copies in PMA. They are:
1. *Bogenträger aus armiertem Beton* (arched girders in reinforced concrete), 25712, Feb. 18, 1902.
2. *Randstein für Strassen* (curbs for streets, precast reinforced concrete curbs), 30961, March 25, 1904.
3. *Bahnschiene mit einer in einen Betonklotz eingebetteten Fahrschiene* (railway made from tracks embedded in concrete blocks), 35413, Dec. 29, 1905.
4. *Fuss für Holzmaste* (base for wooden masts), 36099, Feb. 5, 1906.
5. Hauptpatent: *Konstruktion zur Raumabdeckung* (construction for the covering of space), 46928 (French patent no. 450515), Jan. 20, 1909.
6. Hauptpatent: *Verfahren zur Herstellung von Gewölben* (procedure for the production of arches), 56981 (German patent no. 255778), July 14, 1911.
7. Hauptpatent: *Einlagekörper für Eisenbetondecken* (precast enclosure elements for reinforced concrete decks), 60920, Sept. 24, 1912.
8. Hauptpatent: *Verfahren zur Herstellung von allseitig geschlossenen Hohlsteinen* (fabrication of completely enclosed hollow stone blocks), 65491, Aug. 22, 1913.
9. Hauptpatent: *Silomagazin* (storage silos), 72011, Oct. 9, 1915.

17 See brochure, "Maillart & Cie: Ingenieur & Bauunternehmer, Zürich," 1914, PMA.

18 R. Maillart, "Die Sicherheit der Eisenbetonbauten," *SBZ* 53, no. 9 (Feb. 27, 1909): 120.

19 Hauptpatent: *Konstruktion zur Raumabdeckung*, 46928, Jan. 20, 1909, Maillart & Cie, Zurich.

20 R. Maillart, "Eine schweizerische Ausführungsform der unterzuglosen Decke—Pilz Decke" ("A Swiss-developed Form of Beamless Slab—the Mushroom Slab"), *Schweizerische Ingenieurbauten in Theorie und Praxis, International Kongress für Brückenbau und Hochbau*, Zurich, 1926.

21 Years later the extensive research program at the University of Illinois would follow this same idea for boundary conditions. See M. A. Sozen, and C. P. Siess, "Investigation of Multi-Panel Reinforced Concrete Floor Slabs: Design Methods—Their Evolution and Comparison," *Journal of the American Concrete Institute* 60 (Aug. 1963): 999-1028. This paper reviews the Illinois work and other work back to the beginning of the century, but makes no reference to Maillart.

22 Ernst Stettler, "Reflections on Maillart," *Maillart Papers, Second National Conference on Civil Engineering: History, Heritage, and the Humanities*, Princeton University, Oct. 4-6, 1972, published by the Department of Civil Engineering, Princeton University, 1973, p. 133.

23 A. R. Lord, "A Test of a Flat-Slab Floor in a Reinforced Concrete Building" and "A Discussion of the Basis of Design of Reinforced Concrete Flat Slabs," *Engineering News*, Dec. 22 and 29, 1910 and Jan. 12, 1911.

24 Lord's results appeared in the *Concrete Engineers' Handbook*, by George A. Hool, Nathan C. Johnson and S. C. Hollister, first published in 1918 (New York, McGraw-Hill) and still in print by 1947 when it had its 22nd printing. This well-known compendium presents a substantial portion of Lord's work, including his crucial statement (p. 485): "Even on this extremely conservative basis, the average total moment is only $0.025WL$, equal to $0.035w\ell_1(\ell_2 - qc)^2$, as against the value of $0.09w\ell_1(\ell_2 - qc)^2$ used in the Chicago and ACI codes." Thus, Lord's results were only 39% of those in the codes of 1917 when he wrote that statement. As shown in 1921 by H. M. Westergaard, and W. A. Slater in "Moments and Stresses in Slabs," *Journal of the American Concrete Institute* 17 (1921): 415-538, Lord had failed to consider the complete system of concrete and steel and thus had wrongly interpreted his readings. But the *Handbook* authors stated (p. 487): "The intimate connection which Mr. Lord has had with the development of this branch of engineering and his thorough knowledge of the practical construction side entitles him to speak with authority and with his recommendation we are thoroughly in accord."

Lord recommended the coefficient 0.08 instead of 0.09 in the code because he found only 0.035, but the true value should be 0.125 or over 3½ times the value he got! Although the ACI code never dropped below 0.09 (already low), the *Handbook*, remaining an important work until after Maillart's death, perpetuated these judgments based upon improper test interpretation. The paper by Sozen and Siess (see note 21 above) finally brought this question back for review in a clear and convincing way. The reason Lord got such low values is that at the relatively low loads for which he took measurements, the concrete is able to carry a substantial portion of the tensile stress, all of which the steel is normally designed to take. Thus, it would have been necessary to measure concrete strains as well and to combine values—always a difficult task. Maillart's approach was to derive bending moments from deflections which automatically account for the combined resistance of steel and concrete. His problem was to get enough measurements for accuracy, and his elegant methods merit a fuller description than the present work allows.

25 C.A.P. Turner, *Concrete Steel Construction*, Minneapolis, 1909.

26 Hool, Johnson, and Hollister, *Concrete Engineers' Handbook*, 22nd impression, 1947, p. 480.

27 Early American designs for flat slabs were generally safe enough because designers apparently were guided by field experience and overall load test results, much as Maillart was. Lord's interpretations, however, were definitely on the unsafe side.

Chapter 7

1 M.-C. Blumer-Maillart, "Recollections of My Father," PMA, gives an account of Maillart's return to Switzerland.

2 M. Roš, "Versuche und Erfahrungen an Ausgeführten Eisenbeton—Bauwerken in der Schweiz, 1924-1937," *Bericht, Nr. 99, Beilage zum XXVI Jahresbericht des Vereins schweizerischer Zement- Kalk- und Gipsfabrikanten*, Zurich, 1937, pp. 346-377. (Hereafter *Bericht 99*).

3 David P. Lamb, "The Mushroom Column System of Maillart and the Americans," Senior Thesis, Dept. of Civil Engineering, Princeton University, April 18, 1975, pp. 31-32.

4 R. Maillart, "Ein schweizerische Ausführungsform der unterzuglosen Decke—Pilz-Decke," *International Kongress für Brückenbau und Hochbau*, Zurich, 1926. This volume has only one paper out of twenty-four devoted to buildings, twenty on bridges, and three miscellaneous (a reservoir roof, spherical contact analysis, and a study of airplane wings). Ironically, the only building paper is Maillart's, Switzerland's most famous twentieth-century designer of concrete bridges, while none of the bridge papers are his.

5 E. Bonjour, H. S. Offler, and G. R. Potter, *A Short History of Switzerland*, Oxford, 1952, pp. 292-296.

6 C. Baedeker, *Switzerland*, Leipzig, 1905, pp. 131-132.

7 Such forms have been used frequently for bridges (see, for example, the Aare bridge near Brugg of 1904 in Bühler, "Die Brückenbauten der Schweizerischen Bundesbahnen in den Jahren 1901 bis 1926," *International Kongress für Brückenbau und Hochbau*, Zurich, 1926, p. 51). The connection between flat top and curved bottom is normally by a maze of verticals and diagonals. The idea for a roof system similar to the Gotthard bridges' rebuilt form does appear as early as 1846 in the first major writing of the noted German engineer, Johann Wilhelm Schwedler (see August Hertwig, *Schwedler: 1823-1894*, Berlin, 1930, pp. 40-48 and esp. Table 1, Figure 26). However, Schwedler did not speak specifically of this system, mentioned with forty-seven others in the article dealing with long-span roof systems. Maillart knew Schwedler's works well from Ritter's lectures; see R. Maillart, "Class notes on Graphic Statics of Prof. Wilhelm Ritter," 1892-1893, PMA. Nearly all the second section of these notes is devoted to Schwedler trussworks, and on the page marked "Graphic Stat. II," there appears a truss with a curved top chord which has a kink at the center as in Chiasso.

8 For a detailed description of this rationale, see Robert Mark, James K. Chiu, and John F. Abel, "Stress Analysis of Historic Structures: Maillart's Warehouse at Chiasso," *Technology and Culture* 15, no. 1 (Jan. 1974): 49-63.

9 Ibid. Maillart's design is even more subtle and more rational than my brief discussion might imply. The analysis by Mark, Chiu, and Abel shows, for example, that in the element C_2 the maximum compression occurs on the inside edge, just where Maillart built his transverse frame.

10 In a truss with parallel chords, the maximum compression force in the top is equal to the maximum tension force in the bottom; but for concrete the allowable compression stress would be only about one-twentieth the comparable tensile stress in steel, so that the bottom member needs only enough concrete to cover a steel area one-twentieth that of the concrete area needed in the top member.

11 A. Vierendeel, Livre 2, "Calcul des poutres Vierendeel," *Cours de stabilité des constructions*, Louvain, 1935, pp. 383-386. Vierendeel gives a full bibliography on his girder invention on which the earliest articles were already published by 1898, and the first in German in 1902. Vierendeel's initial publication in French (listed on the back cover of the *Cours*) appears to be dated 1898, although he notes on p. 217 that he developed the theory in 1897.

12 Ibid., p. 217. Vierendeel begins his presentation by stating that "the idea of the girder without diagonals dates from 1896."

13 As an example of this uncertainty about Vierendeel girders, see H. Cross and N. D. Morgan, *Continuous Frames of Reinforced Concrete*, New York, Wiley, 1932, pp. 236-237. "Economy and rigidity are claimed for these structures when built in steel. Any economy that has been realized in such cases seems to be based on the practice of increasing the allowable stress over that used for primary stresses in ordinary steel trusses on the ground that the ordinary truss has secondary stresses in addition to the primary stresses. However, most bridge engineers in America would hesitate to consider the secondary stresses in an ordinary riveted truss of importance equal to the primary bending stresses in a Vierendeel truss."

"The term 'rigidity' represents a property which is ill defined but is usually represented as very desirable. If deflection is meant, it seems evident that the Vierendeel girder is much less rigid."

14 R. Favre, "Die Erneuerung von zwei Maillart-Brücken," *SBZ* 87, no. 17 (April 24, 1969): 315.

15 Calculations for these longitudinal moments due to live load give tensile stresses of about 200-300 psi over the columns where Maillart provided only a very small reinforcement. His detailing of the deck ribs clearly shows the design idea of members fully supported over the columns, whereas in his later deck-stiffened designs the reinforcement is placed in such a way as to ensure that these cracks could not open.

16 Maillart, *Bogenträger aus armiertem Beton*, patent no. 25712, Swiss Patent Office, Bern, Feb. 18, 1902.

17 Ibid.

18 Ibid.

19 James Chiu and David P. Billington, "Geneva Archive of Robert Maillart," Report, Dept. of Civil Engineering, Princeton University, April 16, 1974, pp. 12-17, PMA.

20 Gustav Kruck, "Das Kraftwerk Wäggital," *Neujahrsblatt, Naturforschenden Gesellschaft in Zürich*, no. 127 (1925): 9.

21 Ibid., p. 55.

22 Figure 7-11b is based on Maillart's sketch, PMA. The Vierendeel drawing appears in his "Calcul des poutres Vierendeel," *Cours de stabilité des constructions*, p. 299.

23 Maillart, "4021/88 Statische Berechnung, Eisenbetonbrücke über den Flienglibach," 1923, PMA. "The supporting structure is designed as a stiffened arch. The two parapets form the stiffening members while the relatively very thin arch is used as the supporting line for dead weight and half the traffic, i.e., live load. The moment of inertia [i.e., the stiffness] of the arch with respect to that of the stiffening members (0.007 against 0.318) will be negligible as will the very small arch bending moments resulting from the live loads."

24 Chiu and Billington, "Geneva Archive of Robert Maillart," pp. 16-18. During his reestablishment in Geneva, Maillart made a number of engineering studies, including his well-known shear-center innovation (see E. Reissner, "A History of the Center-of-Shear Concept—Maillart's Works and Ramifications," *Maillart Papers, Second National Conference on Civil Engineering*, pp. 77-96). These studies are listed in Appendix A and show his intense activity between 1921-1924.

25 *SBZ* 58, no. 3 (July 15, 1911): 33.

26 *SBZ* 90, no. 11 (Sept. 10, 1927). One reason the city came to Maillart may have been the strong support he had received for his original 1911 design from the *SBZ* who, in their criticism of the 1911 jury's decision, answered point by point the objections to Maillart's proposal. See *SBZ* 58, no. 3 (July 15, 1911): 33-39, as well as 90, no. 11 (Sept. 10, 1927): 142. Maillart attributed his low position in 1911 to Moser, who was one of the five jurors and apparently the dominant one, because, as Maillart said, "those projects which approach closest to Moser's narrow ideas have won the prizes." See Maillart's letter, April 19, 1911, written to his wife just after the announcement of the awards, PMA. In May, Maillart met with the architect Joss and the editor of the journal *Baukunst* in Bern and agreed that their project would apear there in spite of anticipated objections by Moser. See Maillart's letter of June 1, 1911 to his wife, PMA, and Maillart, "Ansprache," from the opening of the bridge on Saturday, May 17, 1930 and quoted, apparently in full, in the *Berner Tagblatt*, Bern, May 19, 1930.

27 Maillart, "Die Lorraine-Brücke über die Aare in Bern," *SBZ* 97, no. 1 (Jan. 3, 1931): 1-3; no. 2 (Jan. 10, 1931): 17-20; no. 3 (Jan. 17, 1931): 23-26; no. 5 (Jan. 31, 1931): 47-49.

28 Maillart to his daughter, March 26, 1930, PMA.

29 See Maillart brochure, PMA, where, under the Rheinfelden bridge, there is a note identifying the German patent (DRP No. 255778) but not the Swiss one, which he had submitted on July 14, 1911).

Chapter 8

1 M. Roš, "Zur Zerstörung der Rheinbrücke bei Tavanasa," *SBZ* 90, no. 18 (Oct. 29, 1927): 232-236. The high Rhine water accompanying this disaster carried away other works as well; see *SBZ* 90, no. 16: 206-209; no. 17: 223; no. 25: 320-324; and Roš, *Bericht 99*, Vol. 1: 90.

2 M. Roš, "Zur Zerstörung der Rheinbrücke bei Tavanasa," p. 235.

3 M. Roš, "Neuere schweizerische Eisenbeton-Brückentypen," *SBZ* 90, no. 14 (Oct.

1, 1927): 172-177. This article describes the Valtschielbach bridge in detail, briefly gives results from the testing of the Flienglibach and Schrähbach bridges, and concludes with a presentation of the aqueduct at Châtelard.

4 Drawings no. 139 and 140 from the Maillart Zurich office, 1927, PMA.

5 Ganzoni, "Protokoll," Conference in Tavanasa, Dec. 22, 1927, PMA. "The upper Rhine Bridge, 85 meters long and 4 meters in free width, with a roadway above, stands in contrast to a variation from the town of Tavanasa which is about 52 meters long and which has the roadway at the level of the river bank. The lower 47-meter-long secondary road bridge over the Rhine will come to replace the present old wooden bridge."

6 Curiously, in the final cost reckoning for the bridge there were payments of 300 francs given to Maillart & Cie, Geneva, Hartmann & Cie, St. Moritz, and Westermann & Cie, St. Gallen. No mention was made of Prader & Cie. It would appear that these were in the nature of competition prizes. Later on Maillart & Cie was given 1415.30 francs, apparently for the consulting report (see note 8). Another explanation could be that Maillart, Hartmann, and Westermann all did some consulting and that the second and larger payment to Maillart was for his work in the load tests (see note 9).

7 "Protokoll No. 453, von dem Kleinen Rat des Kantons Graubünden," meeting of March 2, 1928, PMA. This bridge was designed by W. Versell of Chur.

8 Maillart, "Bericht über die Fundation der neuen Rheinbrücke bei Tavanasa," Geneva, April 12, 1928, PMA.

9 M. Roš, "Ergebnisse der Belastungsversuche an der neuen Strassenbrücke bei Tavanasa," *Bericht 99*, Vol. 1: 50-55.

10 Mathias Thöny, *Schuders und seine Bewohner*, Schiers, undated but probably shortly after 1956.

11 Mathias Thöny, *Schiers*, Buchdruckerei Thöny, Bruner and Co., Schiers, 1934, pp. 130-131.

12 J. Solca, "Leitsätze für die ausführung der Brücke als Betonbogenbrücke," *Beilage No. 9 Zum Verlag vom 20 Juli 1929*, Chur, July 10, 1928, PMA.

13 Flury, letter from the town council of Schiers to Solca, July 26, 1928, PMA. Two major concerns of the community were employment for local workers and the use of local wood for the scaffold.

14 Prader & Cie, Zurich, to building department of the Canton of Graubünden, Chur, Sept. 15, 1928, PMA.

15 S. Giedion, *Space, Time and Architecture: The Growth of a New Tradition*, 5th ed., 1967, p. 476.

16 Benjamin R. Barber, *The Death of Communal Liberty: A History of Freedom in a Swiss Mountain Canton*, Princeton, N.J.: Princeton University Press, 1974. See especially the chapters, "The Alpine Environment," pp. 79-106, and "The Confrontation with Modernity," pp. 237-274.

17 P. Lorenz to J. Solca, Filisur, Sept. 24, 1928. Lorenz noted that he enclosed the summary, report, and recommendation and made a few further comments on errors in several proposals as well as one in the competition rules. The latter, he stated, would not affect the judgments. The report and recommendations appear together in three typed pages also from Filisur and also dated Sept. 24, 1928, PMA.

18 P. Lorenz, report, Sept. 24, 1928, p. 1.

19 Ibid., p. 2.

20 Ibid.

21 Ibid.

22 Marcel Fornerod, "Reflections on Maillart," *Maillart Papers, Second National Conference on Civil Engineering: History, Heritage and the Humanities*, Princeton University, Oct. 4-6, 1972, pp. 139-140. The objections also included concern over the weathering because at the springing lines water can collect and stay in debris-filled hinge insets. In fact, this did happen at Salginatobel, and after forty-three years the abutment hinges did need extensive repair. The problem, however, is more associated with proper drainage and periodic inspection, than with a basic difficulty of hinges per se.

23 M. Roš, "Bericht zu den Ausführungsplanen der Strassenbrücke über das Salginatobel," sent to J. Solca and dated July 20, 1929, 5 pages, PMA.

24 Maillart's response to Roš's report came in a letter from Prader & Cie to Solca on Aug. 1, 1929, pp. 1-2, PMA. The Pont Candelier was described by Eugène Freyssinet, "Le Pont Candelier," *Annales des ponts et chaussees*, Paris, 1923, pp. 165-197.

25 Roš to Solca, Aug. 9, 1929, PMA.

26 Maillart, private letters, PMA. On February 5, 1929, Maillart's daughter Marie-Claire married the engineer Edouard Blumer, one of Roš's assistants. Shortly thereafter they left for Indonesia where Blumer took on civil engineering work for Royal Dutch Shell. This separation provided the reason for the weekly letters that continued almost to the end of Maillart's life. His first letter was written on March 6, 1929. In the ninth letter, written on June 13, he spoke of having dinner with Roš, and on the eleventh he wrote of visiting the Salginatobel on Saturday (June 22). On July 4 he wrote of just returning again from inspecting the foundations for the Salgina bridge, and on July 9 of attending a big dinner given by Roš in Zurich the past Saturday (July 6). In the letter of July 17 he specifically noted having dinner on Saturday (July 13) in Zurich without "the Professor" (Roš). Again in the letter of July 23 he noted Roš's absence at their previously regular Saturday lunch in Zurich because the "Schutzen" restaurant has gone downhill and "le prof. n'y vient plus." He further remarked that the Salgina was going slowly. Finally, in a letter of August 13, Roš appeared again, having lunch on Saturday noon at the Buffet de la Gare. It is in this letter that Maillart confused Valtschiel with Salgina when he indicated it would be cast in September (see Chapter 9, note 2).

27 Maillart, letter of July 23, 1929. For a brief sketch of the designer, Richard Coray (1869-1946), see *Terra Grischuna, Bündnerland* 33, no. 5 (Oct. 1974): 280-281, PMA.

28 Maillart, letter of Oct. 23, 1929. Here he wrote Valtschiel and crossed it out for Salgina.

29 See letter of May 5, 1931, where in responding to his daughter who had made fun of his cooking, he wrote: "Certainly, if you return [from Indonesia] I will make you a chicken consulting-engineer'! Your are wrong, Picci, to make fun of me, because it is very good in spite of the fact that I have no cookbook. Indeed, it is a bit like with the reinforced concrete. I have bought no books and read no journals in this specialty for 10 years. And even so 'the soup' [the chicken and the concrete] succeeds just about always." Maillart was at the time living alone in the small apartment in his Geneva office.

30 George C. Collins, "The Discovery of Maillart as Artist," *Maillart Papers, Second National Conference on Civil Engineering: History, Heritage, and the Humanities*, Princeton University, Oct. 4-6, 1972, pp. 35-60. The discovery, according to Collins, took place in 1930 and was specifically confirmed by an article by Sigfried Giedion, "Maillart, Constructor of Mushroom Slabs and Their Utilization for Architectural Purposes," *Cahiers d'Art* 5 (1930): 146-152.

Chapter 9

1 In using the word "scientific" here, I am speaking of the uses of scientific ideas within engineering and not their uses within science itself. Although scientists themselves may disagree on the meaning attached to science here, it does nevertheless represent well the engineer's view of it and especially the view that arose in the twentieth century within schools of engineering in the United States. For a clear and convincing historical perspective on this semantic question, as well as its influence on engineering, see Edwin T. Layton, Jr., "American Ideologies of Science and Engineering," *Technology and Culture* 17, no. 4 (Oct. 1976): 688-701.

2 *SBZ* 90, no. 14 (Oct. 1, 1927): 172-177. Maillart himself referred to deck-stiffened arches as bridges of the Valtschiel type, as for example in a letter of July 23, 1930, where he spoke of the Habkern bridge as "tout semblable a celui de Valtschiel." Furthermore, he frequently wrote "Valtschiel" when he clearly meant "Salginatobel," which indicates the importance in his mind of his early work. Letter to Marie-Claire, Aug. 13, 1929: "J'ai assez à faire, mais ce ne sont pas des choses intéressantes sauf le Pont de Valtschiel qu'on commencera à bétonner au mois de Septembre." Letter to Marie-Claire, Oct. 23, 1929: "Le pont de la Lorraine est bientôt terminé ainsi que le cintre du pont de Salgina." Here he had written "Valtschiel" first and then crossed it out. Letter to Marie-Claire, Aug. 5, 1930: "L'essai de Valtschiel est renvoyé au 18 août." In each case the context makes clear he meant the Salgina. Copies of letters in PMA.

3 Maillart, "Statische Berechnung, Brücke über den Valtschielbach," Geneva, April 20, 1925, drawing no. 4094/4/5, page 4, PMA.

4 A somewhat more accurate way to describe this behavior is by considering a two-hinged parabolic arch under a half load, W_A. We know that the moment diagram will be two antisymmetrical parabolas with maximum values $M_A = W_A \ell^2/64$ at each quarterpoint. The stiffening girder, which must follow the deflected shape of the arch (Fig. 9-3), will also deflect antisymmetrically and have maximum moments $M_G = W_G \ell^2/64$ in which W_G = that part of the total live load carried by the girder. The total live load $W = W_G + W_A$, and thus the total moment $M = M_G + M_A$. Girder deflections are easily computed because the half span acts like a simply supported beam whose vertical displacement $D_G = CM_G \ell^2/EI_G$, in which C = a constant; E = the modulus of elasticity; and I_G = the moment of inertia of the stiffening girder. Arch deflections are also easily computed if we assume the arch to be parabolic with a variation in moment of inertia such that $I_S \cos \theta_S = I_A$ = the crown value; and θ_S = the slope angle of the arch axis at any point. The I in Maillart's bridge's variation is not far from this assumption. Then the arch quarter span displacement $D_A = CM_A \ell^2/EI_A$. The close spacing of the cross walls allows us to assume that $D_A = D_G$ and therefore the result is $M_G/I_G = M_A/I_A$. In other words, the ratio of live-load moment in the arch to live-load moment in the girder is the same as the ratio of their respective moments of inertia. Then the arch moment $M_A = M_G I_A/I_G$ or in terms of the total moment $M_A = M/(1 + I_G/I_A)$.

For the Valtschielbach bridge $I_G = 0.1232$ m^4, $I_A = 0.00556$ m^4, and $W = 0.9$ tons/m, so that $M = 0.9 (43.2)^2/64 = 26.2$ mT. Thus, the quarterspan moment in the unstiffened arch would be 26.2 mT, whereas with the stiffened arch its value reduces to $M_A = 26.2/(1 + 22.2) = 0.043 (26.2) = 1.13$ mT with the girder carrying 26.2 − 1.13 = 25.07 mT. That moment in the arch leads to a tension stress of $\sigma = \dfrac{1.13 \times 0.135}{.00556} = 27.4$ T/m^2 = 2.74 kg/cm^2 (39 psi), which is negligible.

5 The normal assumption in arch analysis is that all loads are essentially in the vertical plane of the axis of the arch. In major American works on arches of the early 1930's there is no mention of a vertical loading that is curved in plan. See, for example, H. Cross and N. D. Morgan, *Continuous Frames of Reinforced Concrete*, New York, Wiley, 1932; and C. B. McCullough and E.S.T. Thayer, *Elastic Arch Bridges*, New York, Wiley, 1931.

6 The most famous examples probably are Eiffel's Garabit viaduct in iron and Maillart's Salginatobel bridge in concrete.

7 See, for example, "Brücken und Stege im Bündnerland," *Terra Grischuna, Bündnerland* 33, no. 5 (Oct. 1974), which shows a series of stone bridges with such parapets: p. 253, bridge over the Tuorsbach in Bergün, 1894; p. 255, road bridge by the Solis railway bridge, 1868; and p. 279, road bridge by Lukmanier.

8 R. Maillart, "Aktuelle Fragen des Eisenbetonbaues," *SBZ* 111, no. 1 (Jan. 1, 1938): 1-5. This article, whose title may be translated roughly as "Current Questions in Reinforced Concrete Structures," was taken from a talk given by Maillart to the Schweizerischer Verband Für die Materialprüfungen der Technik on Nov. 12, 1937. Part of it appears in English in Max Bill's *Robert Maillart*, 3rd ed., 1969, pp. 17-18, under the title, "Design in Reinforced Concrete" (where "design" is a translation of *Gestaltung*). The article was far more important than is apparent from that excerpt, however, because it dealt with a fundamental technical question in concrete structures, which Maillart presented in formulas and graphs as well as in the ideas on design which were all that Bill quoted.

9 R. Maillart, "Die Sicherheit der Eisenbetonbauten," *SBZ* 53, no. 9 (Feb. 27, 1909): 119-120. See Chapter 5, note 15 above for a full translation in English.

10 C. T. Morris et al., "Final Report of the Special ASCE Committee on Concrete Arches," *Transactions ASCE*, 100 (1935): paper no. 1922, pp. 1427-1581.

11 Ibid., p. 1429.

12 See, for example, the paper of one member of the five-man committee, A. C. Janni, "The Design of a Multiple-Arch System and Permissible Simplifications," *Transactions ASCE*, 88 (1925): paper no. 1571, pp. 1142-1182, and especially the Appendix, p. 1160, where he referred extensively to European literature, including the works of Maillart's teacher, Wilhelm Ritter. Furthermore, photos of both the Valtschielbach and Klosters bridges had appeared in the article by M. S. Ketchum, Jr., "Thin-Section Concrete Arches as Built in Switzerland," *Engineering News Record*, Jan. 11, 1934, pp. 44-45. This article, published nine months before the committee's report and surely seen by its members, contained a clear description of the structural behavior of deck-stiffened arches: "Investigation of the interaction between arch rib and deck in the reinforced concrete arch by means of models has shown that although the live-load stresses in the rib are considerably reduced by the stiffness of the deck, stresses in the superstructure are increased. At the present time mathematical analysis of these stresses, although possible, is out of the question for ordinary design. To get away from these uncertainties, R. Maillart, consulting engineer, Geneva, Switzerland, has built a number of reinforced concrete arches, using a flexible rib with a stiffening girder. Although relatively unknown in this country, this type of bridge has been used in Europe for more than 40 years, chiefly for bridges of steel. As constructed in reinforced concrete, the arch rib and the posts are thin slabs with little flexural rigidity, while the deck is made very rigid by deep girders. The structural action is similar to that of a suspension bridge. The rib is assumed to take only direct thrust and the stiffening girder the unbalanced moments from live load. In addition to stiffening the rib, the girders carry the floor slab and act as railings to protect the traffic. This type of bridge requires a minimum amount of material."

Ketchum wrote the article on the basis of information sent to him by A. J. Bühler, chief bridge engineer of the Swiss National Railway, and by Maillart himself (Maillart to S. [sic] Ketchum, Sept. 18, 1933, PMA). Probably the first American article on deck-stiffened arches was N. M. Newmark, "Interaction between Rib and Superstructure in Concrete Arch Bridges," *Transactions ASCE*, 103 (1938): 62-80. Newmark actually discussed the Klosters bridge but did not mention Maillart.

13 C. T. Morris et al., "Final Report of the Special ASCE Committee on Concrete Arches," *Transactions ASCE*, 100 (1935): 1431.

14 M. Roš, *Bericht 99*. Of the twenty-three arch bridges studied, eleven were designed by Maillart; of the nine beam bridges, one was by Maillart (the Liesberg railway bridge); and of the fifteen buildings studied, three were by Maillart (all flat-slab buildings). The report had five supplements: *Erste Ergänzung 1938-1939*, with tests on three arch bridges, one girder bridge (the Gründlischwand bridge of Maillart), and one flat-slab building; *Zweite Ergänzung 1940*, with tests to destruction of the footbridge (by Max Greuter & Cie) and the Cement Hall (by Maillart) from the National

Exhibition at Zurich of 1939; *Dritte Ergänzung 1941-1942*, with tests on four beam bridges (second tests on Maillart's two beam bridges previously tested), two arch bridges (one by Maillart at Garstatt), and one article on bridge analysis; *Vierte Ergänzung 1943-1945*, with tests on two arch bridges (one by Maillart over the Töss), one beam bridge, six buildings, six bridges previously tested (including the Maillart bridge at Garstatt), and two analysis papers; and finally the *Fünfte Ergänzung 1947*, with the special title of *Lehre und Nutzen* and containing a summary article by Roš himself and a list of his publications.

15 Abd-el-Aziz El-Arousy, "Studien über das elastische Verhalten von Brückengewölben einschliesslich des Zusammenwirkens mit dem Aufbau," *Mitteilung No. 13 aus dem Institut für Baustatik*, Zurich, ETH, 1942, pp. 1-132.

16 El-Arousy, "Studien über das elastische Verhalten," Vorwort.

17 Ibid., p. 74.

18 El-Arousy, "Studien über das elastische Verhalten," pp. 105-114.

19 Ibid., p. 76. Here he referred to American tests by Wilson, which formed the basis for much of C. T. Morris et al., "Final Report," 1935, as well as to those by Jäger published in the German periodical *Beton und Eisen*, July 1936. No mention is made of any Swiss tests, not even those impressive volumes already published by Roš.

20 El-Arousy, "Studien über das elastische Verhalten," p. 114.

21 Maillart, to Milo Ketchum, Sept. 18, 1933, PMA.

22 El-Arousy, "Studien über das elastische Verhalten," p. 88. This development is the same as that done by Maillart in 1923, given in note 4 above.

23 To study the interaction of arch and deck and its influence on arch stress, we write first the ratio of arch and deck bending moments as derived in note 4.

$$M_G/I_G = M_A/I_A, \tag{1}$$

and since the total half-span live-load moment is

$$M = M_A + M_G, \tag{2}$$

then from (1) and (2),

$$M_A = M/(1 + I_G/I_A). \tag{3}$$

The stresses in a rectangular arch section of depth h_A are

$$\sigma_A = M_A h_A/2I_A, \tag{4}$$

which by putting (3) into (4) gives

$$\sigma_A = M/[2I_G(I_A/I_G h_A + 1/h_A)]. \tag{5}$$

To find how σ_A varies with h_A we can study the behavior of the terms affected by h_A, all enclosed in the parenthesis of the denominator. These terms have a minimum value when their derivative with respect to h_A is set to zero, which gives with $I_A = bh_A{}^3/12$,

$$bh_A/6I_G - 1/h_A{}^2 = 0, \tag{6}$$

$$bh_A{}^3/6I_G = 1, \tag{7}$$

or

$$I_G/I_A = 2. \tag{8}$$

Thus, an increase in h_A leads to an increase in σ_A up to $l_A = l_G/2$ or $h_A = \sqrt[3]{6l_G/b}$ after which any increase in h_A results in a decrease in σ_A. From the perspective of a relatively thin arch, increasing material means increasing stress and hence is unfavorable, while from the viewpoint of a relatively thick arch, decreasing material also means increasing stress and is likewise unfavorable.

Applying this simplified analysis to the Schwandbach, we find that for the live load $W = 1.5$ tons/meter, $\ell = 37.4$ m, $l_A = 0.0028$ m^4 and $l_G = 0.0485$ m^4. Then $M = 32.8$ mT and from eq. (3) $M_A = 0.0545 M = 1.79$ mT, giving a stress in the arch (whose depth h_A is 20 cm and whose width b at the crown is 4.2 m) of about 6.4 kg/cm^2 (91 psi tension at the bottom, compression on top), which is small enough to allow for use of the straight line stress distribution of eq. (4).

Thus, live-load bending is reduced from 32.8 mT for an unstiffened arch to 1.79 mT with the stiffening at Schwandbach, or a reduction to 5.5%. Since $l_A < l_G/2$ for the Schwandbach bridge, any increase in its arch thickness will cause the arch stresses to *increase*, because the influence of M_A is greater than h_A on stress, until $l_A = l_G/2$, after which the stresses decrease. For example, if the arch thickness at Schwandbach increases to 41.2 cm, $l_A = l_G/2$, $M_A = M/3 = 10.92$ mT, and $\sigma = 10.92(.206)/.0245 = 91.9$ T/m$^2 = 9.19$ kg/cm^2 (130 psi). Most American arches of about 37.4 meters' span (122 ft.) have depths of at least 41 cm (16 in.). See McCullough and Thayer, *Elastic Arch Bridges*, Table 5, p. 95, where for a 122-ft.-span open spandrel barrel arch with a rise of 18 ft. (5.5 meters) the crown section has a depth of 2 ft. 3 in. (68.5 cm). Also, standard formulas are given on pp. 360-362, from one of which, for example, we would get 0.00001 (11,000 + ℓ^2) = 2.6 ft. or 66 cm. Therefore, it is not hard to see how the advantages of a very thin arch could be overlooked in American practice. Indeed, if, as was not uncommon, $l_A = l_G$ (which in the above case gives $h_A = 52$ cm), then the stress would be 8.8 kg/cm^2; and thus from that perspective a reduction of the arch to the point where $l_A = l_G/2$ would result in an increase of arch stress and hence appear as a disadvantage.

24 For a suggestive discussion of the relationship between engineering structures and living forms, and especially skeletons of such forms, see D'Arcy Thompson, *On Growth and Form*, abridged ed., Cambridge, Cambridge University Press, 1966 (originally published in 1917, revised 1942). In Chapter 8, entitled "On Forms and Mechanical Efficiency," Thompson described the forms of animal skeletons in terms of bridge structures, drawing heavily on the ideas of graphic statics which, as he notes, were first developed by Carl Culmann of the ETH in Zurich. His argument was that biologists can learn from engineering works and not that engineers can learn from nature. Indeed, the example he gave was taken directly from a remarkable event that occurred in Zurich a few years before Maillart's birth when (Thompson, p. 232) "a great engineer, Professor Culmann of Zurich, to whom by the way we owe the whole modern method of 'graphic statics,' happened (in the year 1866) to come into his colleague Meyer's dissecting-room, where the anatomist was contemplating the section of a bone. The engineer, who had been busy designing a new and powerful crane, saw in a moment that the arrangement of the bony trabeculae was nothing more nor less than a diagram of the lines of stress, or directions of tension and compression, in the loaded structure: in short, that Nature was strengthening the bone in precisely the manner and direction in which strength was required; and he is said to have cried out, 'That's my crane!' "

After describing bone structure in terms of stresses, Thompson went on to describe overall skeleton forms in terms of bridge forms. For example, the cantilever bridge form (exemplified by the Firth of Forth bridge) explained the skeletal form of a horse or a dinosaur through the use of bending moment diagrams: i.e., just the principle developed by Culmann and Ritter and taught to Maillart for visualizing the form of bridges.

25 R. Maillart, "Aktuelle Fragen des Eisenbetonbaues," *SBZ* 111, no. 1 (Jan. 1, 1938): 2.

Chapter 10

1 S. Giedion, *Space, Time and Architecture: The Growth of a New Tradition*, 5th ed., 1967, p. 467.

2 Reyner Banham, *Theory and Design in the First Machine Age*, 1960, pp. 320-330.

3 "Aktuelle Fragen des Eisenbetonbaues," *SBZ* 111, no. 1 (Jan. 1, 1938): 1-5.

4 "Official Documents for the Arch Analyses" in J. Abel, ed., *Background Papers, Second National Conference on Civil Engineering: History, Heritage and the Humanities*, Princeton University, Oct. 4-6, 1972, pp. 103-117; David P. Billington, "Explanation of the Arch Analysis," ibid., pp. 91-102.

5 David P. Billington, "An Example of Structural Art: The Salginatobel Bridge of Robert Maillart," *Journal of the Society of Architectural Historians* 33, no. 1 (March 1974): 61-72.

6 "Salginabrücke," *Hoch- und Tiefbau* 30, no. 12 (March 21, 1931): 106-112 (text in both German and French).

7 M. Roš, "Belastungsversuche an der Strassenbrücke über das Salginatobel bei Schiers," *Bericht 99*, Vol. 1: 110-122.

8 See, for example, David P. Billington, "Meaning in Maillart," *Structures Implicit and Explicit*, VIA, Vol. 2, University of Pennsylvania, 1973, pp. 28-39 and notes on pp. 188-192; and R. Maillart, "The Construction and Aesthetic of Bridges," *The Concrete Way* 7, no. 6 (May-June 1935): 303-309. See note 17 below.

9 Max Bill, *Robert Maillart*, 3rd ed., 1969; and Sigfried Giedion, *Space, Time and Architecture*, 5th ed., 1967, pp. 450-476.

10 The curved slab weight comes directly from the computations published in *Background Papers, Second National Conference on Civil Engineering*, p. 104 (see note 4 above).

11 "Entwurf eines Lehrgerüstes" drawing no. 168/4, PMA. This drawing shows clearly that the thin curved slab, to be cast first on the scaffold, is made up of straight segments.

12 For details on the Sarrasin bridge see Roš, *Bericht 99*, Vol. 1.

13 Elizabeth Mock, *The Architecture of Bridges*, New York, Museum of Modern Art, 1949, p. 7.

14 M. K. Hurd, *Formwork for Concrete*, 2nd ed., Detroit, American Concrete Institute, 1969, p. 5.

15 Christian Menn, "New Bridges," *Maillart Papers, Second National Conference on Civil Engineering: History, Heritage and the Humanities*, Princeton University, Oct. 4-6, 1972, pp. 99-113.

16 M. Cayla, "Les Ponts en béton armé de Laifour et d'Anchamps, sur la Meuse," *Le Génie civil* 55, no. 26 (Dec. 29, 1934): 602-604.

17 R. Maillart, "The Construction and Aesthetic of Bridges," *The Concrete Way* (May-June 1935): 303-309. P. M. Shand translated this article from the original French published in *Le Génie civil*, March 16, 1935 and reprinted in *Bulletin technique de la suisse romande*, April 8, 1939, pp. 85-88, but he omitted references to the French bridges. Shand prefaces his translation by the following: "M. Maillart, the distinguished Swiss engineer who has probably designed and built more concrete bridges than anyone else living or dead, has specially authorized the English translation of this article for republication in *The Concrete Way*. It originally appeared in the famous French engineering review, *Le Génie civil*, having been prompted by an article illustrating and describing the construction of a new concrete bridge over the River Meuse, at Laifour, that was published in the same periodical's issue of December 29, 1934."

18 Maillart, "Construction and Aesthetic of Bridges," *The Concrete Way* (May-June 1935): 303-304.

19 Ibid. The statement "to say nothing of making them satisfy oneself" does not appear in the French version so that its origin is in Shand's translation. I accept it here as a reasonable expression of Maillart's ideas, not just because Shand states that the translation is authorized, but also because it seems consistent with Maillart's personality as revealed in private letters as well as public writings.

20 Ibid., p. 305.

21 Ibid.

22 Ibid., pp. 306-307.

23 M. Roš, *Bericht 99*, Vol. 1, p. 121 for the Salgina load of 10 tons and p. 207 for the Felsegg load of 51.6 tons or 25.8 tons per arch. The design total weights are from the official calculations for each bridge: for Salgina see *Background Papers, Second National Conference on Civil Engineering*, pp. 104-109 and for Felsegg see "738/3/1, Thurbrücke Statische Berechnung," 1933, pp. 12-14, PMA.

24 We could analyze the intervening Rossgraben bridge (1932) in the same way to show its modifications, but they are less significant. Probably because Rossgraben, like Salgina, is on a one-lane country road, Maillart did not feel compelled to rethink his Salgina solution to any great degree. At Rossgraben the only major change was the replacement of solid parapet walls by a slender metal railing, thus accentuating further the arch's thin appearance, and probably recognizing also that the Rossgraben crosses its stream without any ravine drop. It is barely 10 meters above the water, whereas the Salginatobel bridge is 80 meters above the Salgina.

25 Maillart, "Construction and Aesthetic of Bridges," *The Concrete Way*, p. 307. Marcel Fornerod, who did the calculations for the Felsegg Bridge, has described its design in his "Reflections on Maillart," *Maillart Papers, Second National Conference on Civil Engineering*, pp. 137-142. He recalled that at Felsegg Maillart first wanted to use the X-column supports which he later used at Vessy. As Fornerod recalls it, "I suggested using frames hinged at the base and after discussion during several Maillart office visits [Fornerod worked in Maillart's Zurich office] and a few comparative calculations, he agreed to use the haunched cross frames that were finally built" (p. 141).

26 Klett and Hummel, "Die Donaubrücke bei Leipheim im Zuge der Reichsautobahn Stuttgart-München," *Die Bautechnik* 40/41 (Sept. 23, 1938): 521-535.

27 Ibid., p. 524.

28 Ibid., p. 535.

29 R. Maillart, "Ueber Eisenbeton-Brücken mit Rippenbögen unter Mitwikung des Aufbaues," *SBZ* 112, no. 24 (Dec. 10, 1938): 289.

30 Ibid., p. 291n.

31 Ibid., p. 292.

32 G. Collins, "The Discovery of Maillart as Artist," *Maillart Papers, Second National Conference on Civil Engineering*, p. 54.

Index

Note: Individual bridges by Maillart appear in one listing under the entry, bridges. For other bridges, see Bonatz, Paul; Hennebique, François; Menn, Christian; Mörsch, Emil; Sarrasin, Alexandre; and under bridge names.

146 Index

Shand, P. M., 142
Siess, C. P., 135, 136
Silo (by Hennebique), 12, Fig. 2-5, 12
Simonett, 72, 84
siphon, *see under* Maillart, Robert, concrete conduit
Slater, W. A., 136
Solca, S., 131, 138
Sozen, M. A., 135, 136
Speer, Albert, 121
Stalden bridge, *see* Sarrasin, Alexandre
Stettler, Ernst, 74, 135
structural analysis: of buried conduit, 45, 133; of
    deck-stiffened arches, 93-94; and design, 101-103;
    of flat slabs, 56-60; and structural behavior, 92, Fig
    9-3, 93, Fig. 9-8, 103, Fig. 9-9, 104; views of Max Rit-
    ter and Maillart on, 100-103
structural design: and analysis, 101-103; contrast be-
    tween bridges and buildings, 51-53, 57-58, 135; of
    deck-stiffened arches, 94-98; of hollow-box arches,
    19, 21-24; principles of, 45-48, 119-121, 133-134;
    views of Max Ritter and Maillart on, 100-103
Studer, Hans, 128-129
Stüssi, Fritz, 127
Stuttgart, 118

Swiss Federal Technical University, Zurich, *see* ETH
Swiss Society of Cement, Lime, and Gypsum Mfrs.,
    43-44
Swiss Society of Engineers and Architects (SIA),
    Zurich branch, 12, 46-47, 132
Switzerland, 3-4, 10, 19, 24, 43, 44, 45, 47, 60, 63, 72,
    74, 75, 127; attitude of Swiss toward public works,
    33-34; contrast of Swiss and American practice,
    21-22; map, 143

tanks, concrete, *see under* Maillart, Robert
Thatcher, Edwin, 128
Thayer, E.S.T., 139, 141
Thiessen, F. C., 130
Thompson, D'Arcy, 141
Thöny, Matthias, 138
Thur River, 31, 32
trusses, concrete: Vierendeel trusses, 68, 71, 137;
    gabled truss, 64-68, 136
Turneaure, F. E., 134
Turner, C.A.P., 57, 136

United States, 10; Swiss and American practice com-
    pared, 21-22; Culmann and Ritter travel in, 7

Uri, Canton of, 66

Versell, W., 80, 138; new Tavanasa bridge, Fig. 8-5, 80
Vierendeel, A.: concrete trusses, 68, 71, 137
Villanueva (Spain), 60
von Emperger, Fritz, 11, 128, 132

Wagner, Richard, 6
Wayss, G. A., 9, 10, 43, 54, 128; Wildegg bridge, 44,
    127, 132
Wayss and Freytag, 118
Wenner, V., 16, 18, 128, 132
Westergaard, H. M., 136
Westermann, 80, 86, 138. *See also* Froté and Wester-
    mann
Wildegg bridge, *see* Wayss, G. A.
Wilson, 140

Zehnder, O., 134
Zuoz (town), 24, 29, 130
Zurich, 3, 5, 6, 24, 29, 35, 45, 63, 89; Maillart office in,
    89

Library of Congress Cataloging in Publication Data

Billington, David P
   Robert Maillart's bridges.

   Includes index.
   1.   Maillart, Robert, 1872-1940.   2.   Bridges—Switzerland.   3.   Concrete construction.   I.   Title.
TG140.M3B55   624.2'09494   78-70279
ISBN 0-691-08203-0